How
the Other Half
Thinks

OTHER BOOKS BY SHERMAN STEIN

Algebra: A Guided Inquiry
(with Calvin D. Crabill and G. Donald Chakerian)

Calculus and Analytic Geometry, 5th ed.
(with Anthony Barcellos)

*Algebra and Tiling: Homomorphisms
in the Service of Geometry*
(with Sándor Szabó)

Geometry: A Guided Inquiry
(with Cal Crabill and G. Donald Chakerian)

*Strength in Numbers: Discovering the Joy
and Power of Mathematics in Everyday Life*

Mathematics: The Man-Made Universe

Archimedes: What Did He Do Besides Cry Eureka?

How
the Other Half
Thinks

Adventures
in Mathematical Reasoning

Sherman Stein

McGraw-Hill
New York   Chicago   San Francisco
Lisbon   London   Madrid   Mexico City   Milan
New Delhi   San Juan   Seoul   Singapore
Sydney   Toronto

Library of Congress Cataloging-in-Publication Data

Stein, Sherman K.
    How the other half thinks: adventures in mathematical reasoning / Sherman
Stein.
        p.   cm.
    Includes index.
    ISBN 0-07-137339-X
    1. Mathematics—Popular works.   I. Title.
QA93 .S675   2001
510—dc21

                                                                2001018039

McGraw-Hill

A Division of The **McGraw·Hill** Companies

1 2 3 4 5 6 7 8 9 0 AGM/AGM 0 8 7 6 5 4 3 2

ISBN 0-07-140798-7

This is the paperback edition of ISBN 0-07-137339-X.

 This book is printed on recycled, acid-free paper containing a minimum
of 50% recycled de-inked fiber.

Contents

Preface

Occasionally, in some difficult musical compositions there are beautiful, but easy parts—parts so simple a beginner could play them. So it is with mathematics as well.

There are some discoveries in advanced mathematics that do not depend on specialized knowledge, not even on algebra, geometry, or trigonometry. Instead they may involve, at most, a little arithmetic, such as "the sum of two odd numbers is even," and common sense. Each of the eight chapters in this book illustrates this phenomenon. A layperson can understand every step in the reasoning.

One of my purposes in writing this book is to give readers who haven't had the opportunity to see and enjoy real mathematics the chance to appreciate the mathematical way of thinking. I want to reveal not only some of the fascinating discoveries, but, more important, the reasoning behind them.

In that respect, this book differs from most books on mathematics written for the general public. Some present the lives of colorful mathematicians. Others describe important applications of mathematics. Yet others go into mathematical reasoning, but assume that the reader is adept in using algebra.

The thinking in each chapter uses at most only elementary arithmetic, and sometimes not even that. Thus all readers will have the chance to participate in a mathematical experience, to appreciate the beauty of mathematics, and to become familiar with its logical, yet intuitive, style of thinking. This is a book of mathematics, not a book about it.

A word about the chapters that involve random events, such as a tossed penny coming up heads or tails: We will use phrases such as "the odds of getting a heads is 1/2" or "in the long run we expect about 50 percent of the tosses to be heads." The precise meaning of such phrases has been the subject of extensive research by statisticians. We will interpret the phrases intuitively, much as William Feller did in *An Introduction to Probability and Its Applications*. As he put it, "[E]veryone has acquired a feeling for the meaning of statements such as 'the chances are three in five.' Vague as it is, the intuition serves as a background and guide."

Each chapter begins with a simple question about strings made up of two letters, usually *a* and *b*. The opening question may lead to others, which are answered in the course of the chapter. Strings of letters may arise in many ways, for instance, from the wins and losses of a baseball team, the heads and tails of a tossed penny, or the pulses and no-pulses of an electronic stream of data. Some originate in chance events, some in carefully planned arrangements.

Typical of this approach is Chapter 5, which raises the question, "How long a string of *a*'s and *b*'s can you make without repeating any triplet?" This question quickly leads to more general ones, which we then settle with the aid of maps of towns and one-way roads. The answer turns out to be of use in measuring by radar the distance to a planet, in transmitting confidential information and checking the reliability of a computer. Thus, while my primary goal is to illustrate the mathematical way of thinking, if a particular result has applications, so much the better.

I hope this book will help bridge that notorious gap that separates the two cultures: the humanities and the sciences, or should I say the right brain (intuitive, holistic) and the left brain (analytical, numerical). As the chapters will illustrate, mathematics is not restricted to the analytical and numerical; intuition plays a significant role. The alleged gap can be narrowed or completely overcome by anyone, in part because each of us is far from using the full capacity of either side of the brain. To illustrate our human potential, I cite a structural engineer who is an artist, an electrical engineer who is an opera singer, an opera singer who published mathematical research, and a mathematician who publishes short stories.

Other scientists have written books to explain their fields to outsiders, but have necessarily had to omit the mathematics, although it provides the foundation of their theories. The reader must remain a tantalized spectator rather than an involved participant, since the appropriate language for describing the details in much of science is mathematics, whether the subject is the expanding universe, subatomic particles, or chromosomes. Though the broad outline of a scientific theory can be sketched intuitively, when a part of the physical universe is finally understood, its description often looks like a page in a mathematics text.

Still, the nonmathematical reader can go far in understanding mathematical reasoning. This book presents the details that illustrate the mathematical style of thinking, which involves sustained, step-by-step analysis, experiments, and insights. You will turn these pages much more slowly than when reading a novel or a newspaper. It may help to have pencil and paper ready to check claims and carry out experiments.

As I wrote, I kept in mind two types of readers: those who enjoyed mathematics until they were turned off by an unpleasant episode, usually around fifth grade; and mathematics aficionados, who will find much that is new throughout the book.

This book also serves readers who simply want to sharpen their analytical skills. Many careers, such as law and medicine, require extended, precise analysis. Each chapter offers practice in following a sustained and closely argued line of thought. That mathematics can help develop this skill is shown by these two testimonials:

A physician wrote, "The discipline of analytical thought processes [in mathematics] prepared me extremely well for medical school. In medicine one is faced with a problem which must be thoroughly analyzed before a solution can be found. The process is similar to doing mathematics."

A lawyer made the same point, "Although I had no background in law—not even one political science course—I did well at one of the best law schools. I attribute much of my success there to having learned, through the study of mathematics, and, in particular, theorems, how to analyze complicated principles. Lawyers who have studied mathematics can master the legal principles in a way that most others cannot."

I hope you will share my delight in watching as simple, even naïve, questions lead to remarkable solutions and purely theoretical discoveries find unanticipated applications.

—Sherman Stein

Acknowledgments

The broader its intended audience, the harder it is to write the book. For this reason, I made sure that the various drafts of the manuscript were read by several people who well represented the readers I had in mind. Max Massie, a bookman, Larry Snyder, a musician and musicologist, and Joshua Stein, a lawyer, made countless improvements in the exposition. My wife, the poet Hannah Stein, not only repaired sentences and rearranged paragraphs, but kept me human during the month upon month I spent at my desk. The mathematicians George Raney and G. Donald Chakerian substantially clarified the exposition.

My editor at McGraw-Hill, Amy Murphy, meticulously went through the final version, making suggestions that improved every chapter. This is the first time in my experience as an author that an editor has been so involved in the details of making a book as good as it can be.

Both the readers and I are deeply indebted to them.

CHAPTER ONE

The Needle and the Noodle

Georges Buffon (1707–1788) made his reputation with the publication of his multivolume *Natural History, General and Particular*, which brought order to much of what was known about the animal and mineral worlds. In an appendix totally unrelated to natural history, he includes a mainly mathematical work, titled *Essay on Moral Arithmetic*. One of the problems he discusses there concerns a needle dropped at random on a floor furnished with regularly spaced parallel lines.

"I suppose that, in a room where the floor is divided by parallel cracks, one throws a stick into the air. One gambler wagers that the stick will not cross any cracks. The other, on the contrary, wagers that it will cross some of them. What are the odds of winning for each of the gamblers? (One could play this game with a needle or a pin without a head.)"

This is known as the *Buffon Needle problem*.

The Needle

For convenience we will assume that all the lines (cracks) are the same distance apart, namely, the width of the slats, and that the length of the needle is the same as the distance between the cracks. We also assume that these lengths are 1 inch.

Buffon and the two gamblers want to know the likelihood that the needle will miss all the lines and also the likelihood that it will cross a line. Because the needle is not longer than the distance between the lines, it cannot cross two lines. (The case in which the needle lands perpendicular to the lines with its ends just touching the lines occurs so rarely that it will not affect our reasoning.) The typical possibilities are shown here.

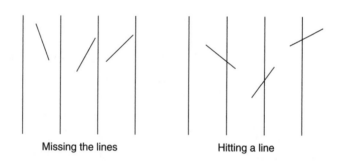

Missing the lines   Hitting a line

We assume that the room is of infinite size. That is, the lines are infinitely long, and there is an infinite number of them. That way, our thinking will not be complicated by a border.

It may seem that the answer to Buffon's question will require a good deal of geometry. However, our reasoning will require none at all, in keeping with my promise in the introduction.

Experiments

It is tempting to guess that the needle will miss the lines as often as it crosses a line. But before we speculate, we should make an experiment. The parallel lines can be supplied by a wood floor or by a floor paved with square tiles. Lacking such a floor, one could draw parallel lines on several pieces of newsprint taped together. A piece of wire as long as the distance between the lines can serve as the needle.

When carrying out the experiment, give the needle a good spin so that it doesn't always fall at the same angle. Also, to help achieve randomness, change the direction in which you stand.

I tossed the needle 100 times. There were 66 cases in which the needle crossed a line and 34 in which it did not. That is pretty far from the guess that the 2 cases would be split evenly. The results of the first 50 of these throws are recorded in the following string of *h*'s and *m*'s, where an *h* stands for hitting a line and an *m* for a miss:

hhhmhmmhmhhmhmhhmhhhmhmhh
mmhmhmhmhhhmmhhmhhhhmhmhhh

There are 31 hits and 19 misses. Of the next 50, 35 are hits and 15 are misses.

But what are the exact odds? I mean, as we toss the needle many times and the string of *h*'s and *m*'s gets longer, what will happen to the percent that are *h*'s?

I find it strange that that percent does tend to settle down and stay closer and closer to some number. After all, the needle has no memory. Each toss is totally independent of all the earlier tosses, yet the percent of hits tends to stabilize as though the needle does remember and wants to hit a line in the long run a certain fraction of the time. We will find that fraction.

Rewording the Problem

On any toss, the needle hits either no lines or one line. To put it another way, there are either no crossings or one crossing. Thinking in terms of crossings, we may ask, "What is the average number of crossings when a needle is tossed billions of times?"

In my experiment there were 66 crossings in 100 trials. That is an average of 66/100, or 0.66 crossings per trial. Therefore, we expect our theoretical average, the one for billions of throws, to be somewhere near 0.66. Whatever the answer is for crossings, it will also tell us the likelihood of the needle's hitting a line.

The advantage of this version in terms of crossings, introduced by Émile Barbier (1839–1889) in 1870, is that it easily generalizes to other geometric shapes. We now ask, "If we have a thin wire of any shape and length, what will be the average number of crossings of the lines when we throw it billions of times?" It turns out, as we will soon see, that this more general question can easily be settled by elementary means. That the more general case turns out to be easier than the specific case is not unusual in mathematics and the sciences. The key to finding the answer may lie in asking the right question. The correct question may offer a clue to its own answer.

The Noodle

Consider any rigid wire made up of straight pieces welded together. The wire must be "flat" in the sense that when it falls on the floor, all of it touches the floor. Here are some of the possible shapes.

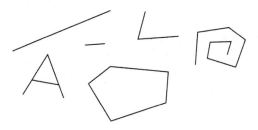

The wire could be straight and of any length, a letter of the alphabet, a spiral, or whatever comes to mind.

We will consider only wires in the shape of polygons. A *polygon* is a figure made of straight segments. In classical geometry a polygon forms a closed circuit, but we will use the term more generally.

We now ask a far more general question than the one about the needle: Handed a flat piece of wire, made of straight pieces and of a certain length and shape, we ask, "How can we predict the average number of crossings when we throw the wire many, many times?" In a 1969 paper on this question, J. F. Ramaley called this the "Buffon Noodle" problem.

For instance, the Z-shaped wire shown to the right can have 0, 1, 2, or 3 crossings. We disregard the rare case when a segment happens to lie on one of the lines. The average number of crossings must lie somewhere between 0 and 3.

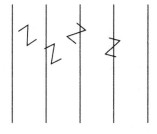

This time it is hard even to guess the answer, for the average depends on the particular wire. The only case in which we have any data at this point is that of the needle, which is as long as the slats are wide.

We restrict our study to polygons—figures made up of straight-line segments—to simplify the mathematics. However, any reasonably smooth curve can be approximated by polygons made up of very short pieces, even pieces all of the same length. Because of this, our theory applies even to curves. In fact, it even applies to polygons and curves made of flexible string instead of rigid wires. For this reason we may speak of Buffon's problem for a wet noodle.

Though we seek a theoretical answer, experiments will serve as a check and may even suggest a solution. We will start with the simplest cases, a common tactic of mathematicians, almost the opposite of another tactic, illustrated in this chapter, which is to generalize.

Experiments

To describe a particular wire, we will use its length and shape. We will assume that the lines are an inch apart. That way, we can easily measure lengths with an ordinary measuring tape used in sewing.

Let us begin with a needle twice as long as the needle we started with. It is 2 inches long and thus can have 0, 1, or 2 crossings. Recall that as we throw it, we give it a good horizontal spin in order to help make the throw random.

Imagine that a bug rides the needle—a bug that will help us in later chapters as well. When the needle lands, the bug crawls from one end to the other and reports how often he crosses a line.

I threw this needle 20 times, with the number of crossings shown here in detail in the order they occurred: 1 2 2 0 2 0 2 1 2 2 2 2 1 0 1 1 2 1 2 2. There were three 0s, six 1s, and eleven 2s. So the average number of crossings per throw, that is, the total number of crossings divided by 20, is

$$\frac{3 \times 0 + 6 \times 1 + 11 \times 2}{20}$$

That is, 28/20 = 1.4. On average, then, the straight wire that is 2 inches long had 1.4 crossings per throw in this experiment.

Then I bent this same wire into the shape of a V with arms of equal lengths. Here are the numbers that the bug reported for 20 throws: 2 1 2 1 2 1 2 1 1 2 1 2 1 1 2 1 1 0 1 1. This time there were one 0, twelve 1s, and seven 2s, for a total of 26 crossings. That is an average of 1.3 crossings per throw.

Next I bent the same wire into a square. The number of crossings for the 20 throws was 0 2 2 2 0 2 2 2 2 2 0 2 0 0 2 2 0 0 2. (This time there could not be any 1s because when a square crosses a line, it crosses it twice. We disregard the rare case when the square just touches a line.) Since there were 26 crossings, the average per throw is 26/20 = 1.3.

Recall the 100 throws of the 1-inch-long needle, during which there were 66 crossings. In this case we see that the average was 66/100, which is 0.66 crossings per throw.

I went on to bend the same 1-inch wire into a Z. Now there could be anywhere from 0 to 3 crossings in a throw. Here are the bug's reports on 20 throws: 3 0 0 1 0 1 0 0 0 1 1 0 0 1 0 0 1 1 2 1. This gives a total of 13 crossings in 20 throws for an average of 13/20 = 0.65 crossings per throw.

All these averages are only suggestive. They are based on only a few throws, not on thousands. However, they may guide us as we frame a theory that does not depend on any experiments.

Interpreting the Data

All our experiments are summarized in this table:

Shape	Length*	Average Number of Crossings per Throw
Straight	1	0.66
Straight	2	1.40
V	2	1.30
Z	1	0.65
Square	2	1.30

*Length of wire.

You could easily add to the list, using longer wires and other shapes. The only two factors that can influence the average are shape and length. Looking at the data, skimpy though they may be, we are tempted to say that shape has little or even no influence on the average number of crossings. For instance, changing the 2-inch wire from straight to a V and then to a square seems not to affect the average significantly. Changing the length, however, exerts a large influence. Let us take a look at the role of shape first.

The Influence of Shape

Let us use common sense to compare the two cases of a wire of length 2, straight and bent into a V. The experimental averages for these shapes were close to each other, 1.40 and 1.30.

Once again we summon bugs to assist us.

First consider the straight needle. Instead of one bug reporting crossings, let us have two bugs. Each bug wanders over his half of the needle, as shown below.

One bug reports crossings of the left section of the needle, and the other bug reports crossings of the right section. Each will report a 0 or 1 because each section of the needle is as long as the width of one slat. To learn the total number of crossings of the whole needle, we listen to the two bugs, and we add the two numbers they utter.

Neither bug knows about the other bug. Indeed, each knows only his own section and is not aware that there is another section. Each bug thinks that his section is being spun and thrown at random.

The average number of crossings for the whole needle is the sum of the averages the two bugs report. Keep this in mind as we now bend the needle into a V.

The wire, which had been straight, is now a V. The two bugs have no idea that they now ride on a bent wire. After all, each of them is aware only of his own section. Moreover, each is responsible only for reporting crossings of his section.

As the V-shaped wire is thrown at random, each of its two sections is also thrown at random. As far as the bugs are concerned, their sections are being thrown just as they were when the wire was straight. Therefore, each bug reports the number of crossings now as it did before, when it was riding half the straight needle. Therefore, the average number of crossings for each bug observed is the same as before. It follows that the (theoretical) average number of crossings for the V is the same as for the straight needle of the same length. Thus we conclude that bending the straight wire into a V of two equal arms has no effect on the average number of crossings if the wire is thrown not just 20 times but millions of times.

However, we don't know at this point what that average is. The experiments suggest that it is somewhere in the vicinity of 1.30 and 1.40.

Our analysis applies to any wire that is bent into the shape of a polygon. As a reminder, here are a few polygons, some of which we have already looked at.

To analyze the Z-shaped polygon, we call upon the service of three bugs. Place one on each of the three sides. (The sides don't have to be the same length.) Each bug has no idea that he is riding on a Z-shaped wire. He thinks he is crawling about on a much shorter straight wire and must report that section's crossings of the lines in the floor.

Now straighten the Z without dislodging the bugs, and don't even tell the bugs what you are doing. The three bugs now report on the three sections of a straight wire. They report the same number of crossings, on average, as they did when they were crawling on the Z. The figure below contrasts the "before" and "after."

When the whole Z-shaped wire is thrown at random, so is each of its three sections. The average number of crossings for the Z is therefore the same as for the straight wire of the same length. This argument applies to any polygon. Just plant a bug on each of its segments and reason as we did for three bugs. We can therefore say that *shape has no influence*

on the average number of crossings of the wire and the lines in the floor.

The Influence of Length

Now that we have ruled out shape as an influence on the average, we are left only with length as a factor. Let us see how the average behaves as we change the length of the wire. We might as well assume that the wire is straight since that case is easiest to draw.

The straight wire of length 2 inches has an average of 1.40 crossings, which happens to be about twice the average for the wire of length 1. That comparison is based just on experiments. What do the bugs tell us about the comparison if billions of throws are made?

Imagine two bugs on the 2-inch-long wire. Each is responsible for half the wire, as shown here.

Once again each bug is unaware of the other bug and the other section. Each reports approximately the same total number of crossings when the wire is thrown billions of times. Therefore the total number of crossings for the whole wire of length 2 will tend to be twice the total number for the wire of length 1. *Doubling the length will double the average number of crossings.*

What if one wire is, say, 2/3 as long as another wire? In this case, divide the longer wire into three sections of equal lengths and place a bug on each section, like this.

The part of the wire, AC, is 2/3 as long as the whole wire, AB. Since there are two bugs on AC and three bugs on AB, the number of crossings of the part AC will tend to be 2/3 the number of crossings of the whole wire. This tells us that the average number of crossings for the shorter wire is 2/3 the average for the whole wire.

This type of reasoning shows that the average number of crossings for a long wire is greater than that for a short wire. Moreover, the average is proportional to the length. Another way to say this is *"If, for a wire, you divide the average number of crossings by the length of the wire, you will get a number, and this number is the same for all wires."*

Let us check that claim for the five wires whose data we recorded in a table:

Shape	Length	Average	Average Divided by Length
Straight	1	0.66	0.66/1 = 0.66
Straight	2	1.40	1.40/2 = 0.70
V	2	1.30	1.30/2 = 0.65
Z	1	0.65	0.65/1 = 0.65
Square	2	1.30	1.30/2 = 0.65

The Missing Number

Thanks to the bugs, we know that for any wire, if we divide the average number of crossings by the length of the wire, we should get the same number as we would for any other wire. This is in fact the case for any curve or polygon. The preceding table suggests that the average divided by the length is somewhere near the range 0.65 to 0.70. But what is this missing number exactly, which is the key to our whole study of crossings? What is this "universal constant" that is not affected by shape or length?

Since the only "famous" number that is near the experimental results is 2/3, we may be tempted to guess that the missing constant is 2/3. As we will see in a moment, that guess is wrong.

Cutting more wires, bending them into various shapes, and then tossing them on the floor, even were we to throw them trillions of times, will not help us find this constant. That would give us only better estimates. Instead, to find that missing constant—the key to the whole chapter, the number that will answer Buffon's original question about a needle—we will have to figure it out by common sense, by pure thought.

The Constant Found

If we could find the constant for just one wire, we would then know it for all possible wires. Which wire will best serve our purposes? Surely not the needle, for we have no way to figure out its average number of crossings any better than we already have with our experiments.

Luckily, there is a certain wire whose average number of crossings we can work out without throwing it even once. That wire is a circle of diameter 1, the same as the width of the slats, the distance between the parallel lines on the floor.

First, we know the circumference (length) of a circle. It is π times the diameter. In our case the diameter is 1, so the circumference is π, whose decimal begins 3.141592. (To remember the digits, say "How I wish I could recollect pi." The length of each word gives a digit.)

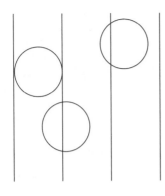

The average number of crossings for this circular wire is easy to find, as shown here.

No matter where the wire lands on the floor, there will always be two crossings. (We count "just touching" as a crossing. After all, the conscientious bug would report it.) So the average must be 2. For this particular shape, then, we have an average of 2 and a length of π. When we divide the average by the length, we do not get 2/3 as our missing universal constant, but 2/π, or 2/(3.141592), which is just under 0.64. That is just a little less than our experimental results, which were in the 0.65 to 0.70 range.

Back to Buffon's Needle

Buffon's original question was not about the average number of crossings. Rather, it asked, "What fraction of the times that you toss a needle, whose length is the distance between lines, does it hit a line?" However, we noticed that that is the same as the average number of crossings for this particular wire. Since the length of the needle is 1, we have

$$\frac{\text{Average}}{1} = \frac{2}{\pi}$$

This equation tells us that the average is 2/π, or about 0.64. That means that the needle will land on a line about 64 percent of the time.

It's rather strange that we used a circle to answer a question about a straight object, the needle. But we could also go in the reverse direction, using the straight needle to find information about the circle. We could throw the needle thousands of times and compute the fraction of times it crossed a line. That would give an estimate of the fraction $2/\pi$, from which we would be able to obtain an estimate of π itself.

Buffon answered his question by the use of calculus, which was invented near the end of the seventeenth century by Newton and Leibnitz. The more elementary solution presented in this chapter exploits the link between length and crossings.

This problem introduced the field now known as *geometric probability*, which combines geometry and probability theory. Moreover, before this problem was posed, probability theory was concerned only with situations with a discrete set of possible outcomes (for instance, the likelihood of getting various totals when two dice are thrown). In contrast, a needle can occupy a continuous array of positions relative to the lines in the floor.

An Application

I would like to include one discovery that is easily established with the aid of our result about the average number of crossings.

Consider a rigid wire in the shape of a convex loop, that is, a loop without dents. (The formal definition is that whenever two points lie in the region bounded by the wire, so does the whole line segment that joins them.) When such a wire crosses a line, it usually crosses it exactly twice.

The loop casts a shadow in every possible direction, as illustrated here, which displays just three of its shadows.

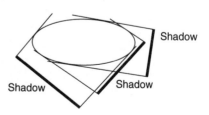

It has been shown that *the length of the curve is equal to π times the average shadow length*. This is quite a generalization of the relation between the circumference and diameter of a circle, a case in which all the shadows have the same length, namely, the length of the diameter.

However, by counting crossings, we will see why the average shadow length is related to the length of the wire.

For convenience we start again with parallel lines 1 inch apart and a wire so small that wherever it falls it never crosses two lines. The wire can fall at any angle with respect to the lines. Let us focus our attention on the case of just one typical angle, as illustrated below. Throughout our discussion, we keep this angle fixed.

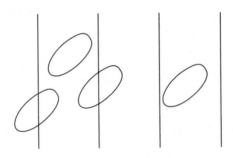

The wire crosses a line whenever its shadow perpendicular to the lines crosses a line, as shown below.

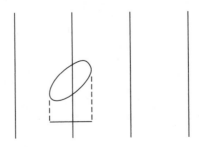

This relation is the key to our reasoning that follows. When the shadow crosses a line, the wire has two crossings with that line. The longer the shadow, the more likely it is that the wire crosses a line. To find out how likely, consider the next figure, in which the width of the shaded band is the same as the length of the shadow.

The shadow meets a line when its left end lies in a shaded band, as shown here.

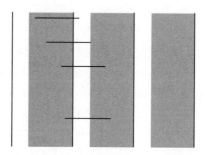

The likelihood of this happening is simply the width of the band since the lines are all 1 inch apart.

Consequently, the likelihood that the wire, falling at the given angle, hits a line is just the length of its shadow perpendicular to the lines. When it does hit a line, there are two crossings. Thus the average number of crossings, when the wire falls in any given angle, is

2 times the length of the shadow perpendicular to the lines

That is the key to our analysis.

Since the wire can fall at any angle, the average number of crossings is

2 times the average shadow length

Now, recall that

$$\frac{\text{Average number of crossings}}{\text{Length of wire}} = \frac{2}{\pi}$$

Therefore,

$$\frac{2 \text{ times average shadow length}}{\text{Length of wire}} = \frac{2}{\pi}$$

Dividing both numerators by 2 gives us

$$\frac{\text{Average shadow length}}{\text{Length of wire}} = \frac{1}{\pi}$$

This last equation tells us that the wire is π times as long as its average shadow. That is what we wanted to show.

The argument we used for short wires can easily be tweaked to apply to any convex wire. Just place the lines far enough apart so that the wire cannot cross two lines. Then use the distance between those lines as the unit of measuring lengths. Incidentally, the relation between the length of a convex wire and its average shadow was established by the French mathematician Augustin Cauchy (1789–1857) in 1841, using calculus.

In 1977 a conference of some 15 mathematicians and 15 biologists was held in Paris to mark the 200th anniversary of

Buffon's problem. The title of the meeting was "Buffon Bicentenary Symposium on Geometrical Probability, Image Analysis, Mathematical Stereology, and Their Relevance to the Determination of Biological Structures." As an indication of the variety of the research, one paper described work related to "forestry sample surveys." The report of the proceedings opened with a reproduction of Buffon's own diagram of his needle problem.

Clearly Buffon's problem has repercussions to this day.

CHAPTER TWO

Win by Two

In the game of volleyball the winning team must have a certain number of points, for instance, 25 in international men's contests. They must also have a margin of at least 2 over the losing team. This means that if the score is 25 to 24, the game is not over. It also tells us that the score just before must have been tied at 24 to 24.

Let's take a look at a game in which the score is tied at 24 to 24. Evidently the two teams are evenly matched. Presumably they have equal odds of winning any particular point. In that, they resemble, perhaps, a tossed penny that can come up either heads or tails. Later we will take a second look at this assumption.

The game goes on until one team leads by 2 points. In theory the game could continue for hours, perhaps forever. If the teams split the next 2 points, the score would then be 25 to 25. Then they could split the following 2 points, the score reaching 26 to 26. Perhaps one team would then win the next 2 points and emerge victorious, with the score 28 to 26.

Close games like these can end with scores of 26 to 24, 27 to 25, 28 to 26, and so on. This scoring situation raises two questions: In such games, how many points do the teams have to play on the average before a game ends? What is the most common final score?

We will approach these questions three ways. First we will examine the records of past volleyball matches, focusing our attention only on ones that were tied at 24 to 24.

Second, we will flip a penny to simulate the points won or lost by the two teams. Heads will mean one team wins, and tails that the other team wins.

Third, we will use just common sense, answering the questions without appealing to volleyball records or a flipped penny. That doesn't mean that our first two efforts will be wasted. They serve as checks on our common-sense approach, as we will see.

If the two teams are the Atlanta Aphids and the Baltimore Beetles, we will let an *a* stand for a point for Atlanta and a *b* for a point by Baltimore. Thus we can record the game as a string of *a*'s and *b*'s.

In the rest of the chapter we will be interested only in the points played after the 24-to-24 score. For instance, the string *aa* tells us that Atlanta won the next 2 points and the game was over at 26 to 24. The string *abbaaa* tells us that the two teams split the next 2 points (and were tied at 25 to 25) and the next 2 (and were tied at 26 to 26), and then Atlanta won 2 points in a row, emerging victorious with the final score of 28 to 26.

The shortest possible string at the end of such a game has length 2, either *aa* or *bb*. The next shortest string has length 4, such as *babb*. The length of a string of interest to us has any of the lengths 2, 4, 6, . . . , in short, any even number. (For instance, say that the game ends with the Baltimore Beetles winning. Then there will be two more b's than a's. Therefore, the number of a's and the number of b's will both be even or both be odd. In either case their sum is even. The same holds if the Atlanta Aphids win.)

The two questions now read: What is the average length of the strings that come from close games? What is their most common length?

Volleyball Records

First let's take a look at what happened in some actual volleyball games. During the 1999 and 2000 seasons, the U.S. national men's volleyball team played 64 games that were tied at 24 to 24. Here is how the final scores turned out in these cases:

Score	Number of Games
26–24	24
27–25	12
28–26	4
29–27	4
30–28	7
31–29	4
32–30	3
33–31	1
34–32	0
35–33	0
36–34	2
37–35	0
38–36	3

Remember that we are interested only in points made after the score was tied at 24 to 24. (The earlier part of the game does not concern us.) Thus for us a score of 26 to 24 means that the score after the 24-to-24 tie was 2 to 0. A 27-to-25 game means a score of 3 to 1 after the tie. With this perspective, we replace the preceding table with the following one:

Score*	Number of Games
2–0	24
3–1	12
4–2	4
5–3	4
6–4	7
7–5	4
8–6	3
9–7	1
10–8	0
11–9	0
12–10	2
13–11	0
14–12	3

*After tie.

To simplify matters further, we record just the total number of points made after the game had reached 24 to 24. For instance, the total in a 2-to-0 game is 2; in a 3-to-1 game, it is 4; and so on. If we know the total number of points, we can figure out the number of points each team scored. For instance, if the total is 8, then the winner had 5 points and the loser 3. The total has to be even since it is either the sum of two even numbers or else the sum of two odd numbers:

Total Points	Number of Games
2	24
4	12
6	4
8	4
10	7
12	4
14	3
16	1
18	0
20	0
22	2
24	0
26	3

The most frequent number of points played after the game had reached 24 to 24 is 2. That is as quick as can be: One team wins both points, the other none. Next most frequent is the case in which 4 points are played. The teams split the first 2 points, then one team goes on to win the next 2 points. The data suggest that shorter games are more frequent than longer ones.

To find the *average* length of a game, we add up the total points played and divide that sum by the number of games, which is 64. This sum would have twenty-four 2s, twelve 4s, and so on. Thus the average number of points scored after the 24-to-24 stage is the following fraction:

$$\frac{\begin{array}{l}(24 \times 2) + (12 \times 4) + (4 \times 6) + (4 \times 8) + (7 \times 10) \\ + (4 \times 12) + (3 \times 14) + (1 \times 16) + (0 \times 18) \\ + (0 \times 20) + (2 \times 22) + (0 \times 24) + (3 \times 26)\end{array}}{64}$$

This equals

$$\frac{48 + 48 + 24 + 32 + 70 + 48 + 42 + 16 + 0 + 0 + 44 + 0 + 78}{64}$$

or

$$\frac{450}{64}$$

This fraction equals about 7.03. For those 64 games, the average number of points played after the 24-to-24 tie was therefore about 7.

Now let us see what a penny tells us.

The Penny

Since the teams were tied at the end of 48 points at 24 to 24, it is reasonable to assume that they are equally matched. Therefore, we will assume that each team has the same chance of winning any rally. That means each team has a 50 percent chance, or 1/2, of winning any particular point. For this reason we suspect that their play should resemble the fate of a tossed penny that is equally likely to come up heads or tails. We interpret heads as a point for Atlanta and tails as a point for Baltimore. Later we will see that the assumption that the two teams behave like tossed pennies is only a first assumption, and it has to be modified.

This comparison suggests that we toss a penny to imitate two teams playing volleyball. We flip it until the total number of heads exceeds the total number of tails by 2, or the total number of tails exceeds the total number of heads by 2. We then record the total number of tosses, which could be as few as 2. However, just as with a volleyball game, we may have to toss the penny many times.

Here is a record of a penny experiment that needed 10 tosses:

ab|ab|ba|ab|bb

An *a* stands for heads, a *b*, for tails. I put the vertical bars in to make it easier to see at a glance what happened. The number of heads equaled the number of tails at the end of 2, 4, 6, and 8 throws. Then 2 tails showed up, and tails obtained a lead of 2. The "score" then is 6 to 4. This corresponds to a volleyball game during which 10 points were played after the 24-to-24 tie and the total score at the end was 30 to 28.

I performed the experiment 64 times, the same as the number of volleyball games. In each experiment I tossed the penny until the number of heads differed from the number of tails by 2. The following table records the results:

Total Number	Number of Times
2	31
4	14
6	8
8	3
10	2
12	3
14	1
16	0
18	1
20	1

In both cases, the volleyball and the penny, the most frequent total is 2, the next most frequent is 4, then 6. Volleyball has more long runs than the penny. Because of these long runs, the average for volleyball is probably larger than the average for pennies. The penny's average run is the following fraction:

$$\frac{\begin{matrix}(31 \times 2) + (14 \times 4) + (8 \times 6) + (3 \times 8) + (2 \times 10)\\ + (3 \times 12) + (1 \times 14) + (0 \times 16) + (1 \times 18) + (1 \times 20)\end{matrix}}{64}$$

This equals

$$\frac{62 + 56 + 48 + 24 + 20 + 36 + 14 + 0 + 18 + 20}{64}$$

which is

$$\frac{298}{64}$$

This equals about 4.66, lower than the 7.03 average for the volleyball games. That discrepancy is so large that it hardly could be due to chance. Later we will have an explanation from volleyball experts for the difference.

You are invited to perform your own experiments with a coin of your choice and compare the results with the ones described.

An Approach without Experiments

To make better estimates, we could either collect more data on volleyball tournaments or throw a penny many times, perhaps thousands of times. If we had a computer with a random-number generator, we could run off a million cases while we slept. In any case, the more we experiment, the more reliable our estimates will be.

There is, however, a method for predicting the frequencies of the various scores without doing any experiments whatsoever. In this approach we combine our common sense with a little arithmetic. The experiments will not only guide us but, when we are done, serve as a check on our thinking.

To begin, look closely at the fraction that represents the average for the 64 penny trials:

$$\frac{\begin{array}{l} (31 \times 2) \;+\; (14 \times 4) \;+\; (8 \times 6) \;+\; (3 \times 8) \;+\; (2 \times 10) \\ +\; (3 \times 12) \;+\; (1 \times 14) \;+\; (0 \times 16) \;+\; (1 \times 18) \;+\; (1 \times 20) \end{array}}{64}$$

Dividing each term by 64, we rewrite this as the sum of several fractions:

$$\left(\frac{31}{64} \times 2\right) + \left(\frac{14}{64} \times 4\right) + \left(\frac{8}{64} \times 6\right) + \left(\frac{3}{64} \times 8\right) + \left(\frac{2}{64} \times 10\right)$$
$$+ \left(\frac{3}{64} \times 12\right) + \left(\frac{1}{64} \times 14\right) + \left(\frac{0}{64} \times 16\right) + \left(\frac{1}{64} \times 18\right) + \left(\frac{1}{64} \times 20\right)$$

This expression is the sum of 10 quantities. Each is a fraction times an even whole number.

What do the fractions mean? The first, 31/64, is the fraction of the 64 experiments that had a total of 2. The second is the fraction of times that the total was 4, and so on. If we performed billions of experiments, what would happen to these fractions? In other words, using just common sense, how would we expect those fractions to behave? Would they wander all over, or would they tend to get closer and closer to some fixed fraction? If the latter happens, what would we expect that fixed fraction to be?

Let us now think this through. Imagine tossing a penny. In the first two tosses any of four outcomes can occur:

aa (heads, heads)   *ab* (heads, tails)
ba (tails, heads)   *bb* (tails, tails)

These four possibilities are all equally likely since the penny is just as likely to come up heads as tails. Two cases—namely, *aa* and *bb*—result in a difference of 2 between heads and tails. So we would expect that on average two out of four trials would end with a score of 2 to 0. On the other hand, the cases *ab* and *ba* result in ties at 1 to 1.

Imagine that there were 1000 games that were tied at 24 to 24. We would expect about half of them, that is, 500, to be over after just 2 more points are played. Also 500 would be tied at 25 to 25. Then 2 more points are played. In half of these cases one team takes a 2-point lead and the game is over with a score of 27 to 25. In 250 games the score is again tied.

We see that in 1000 close games, there would be around 250 that would end with a score of 27 to 25. That is a quarter of the games. In terms of the penny, we see that about one quarter of the experiments would end with a total of 4 tosses. As a check, let us compare this conclusion with the data.

There were 14 cases out of 64 in which the total was 4. That is very close to our theoretical 1/4. As percents, 14/64 is about 22 percent, while 1/4 is exactly 25 percent.

We expect half of the trials to end after just 2 tosses and a quarter to end with 4 tosses. So three quarters of the trials end with a total of 2 or 4 tosses. It follows that one quarter will go on longer.

Which ones go on longer? Those in which the first 2 tosses are opposites, a head and a tail in one order or the other, and the next 2 tosses are also opposites. This occurs in 1/4 of the trials. In half of these cases the next 2 tosses will be *aa* or *bb,* resulting in a lead of 2 at 6 tosses. Since half of 1/4 is 1/8, we expect a total of 6 to occur in one eighth, or 12.5 percent, of the trials. In our experiment the observed fraction was 8/64, which is exactly 1/8, the predicted value.

Continuing this reasoning tells us that the fraction of trials that end with 8 tosses is 1/16. The theoretical fractions so far are 1/2, 1/4, 1/8, 1/16. The pattern goes on this way, each fraction being half the preceding one.

Here is one way to see why this is so. In a trial, group the tosses into pairs, as indicated in this diagram, which shows a typical case of a lead of 2 arising after 10 tosses:

Each dash would hold an *a* or a *b*. The vertical bars separate the tosses into pairs.

In the first block of 2 there must be a tie, one *a* and one *b*. That happens half the time. The same must occur in the next block of 2. So in half the cases in which there was a tie in the first block, there is also a tie in the second block. So in half of a half, or a quarter, of the trials, there is a tie at the end of 4 tosses. In half of these cases there will be another tie in the next block. So in one eighth of the times there is a tie at the end of 6 tosses. Similarly, at the end of 8 tosses, there is a tie in one sixteenth of the cases. But in the fifth block of 2 either 2 heads or 2 tails appear. That occurs in half the cases in which there was a tie at the end of 8 tosses. All told, then, a total of 10 should occur on average about once in 32 trials. That's about 3 percent; we had 2 in 64 trials, which is in the same proportion.

It follows from this line of reasoning that the average number of tosses until there is a lead of 2 (if we threw the penny billions of times) is

$$\left(\frac{1}{2} \times 2\right) + \left(\frac{1}{4} \times 4\right) + \left(\frac{1}{8} \times 6\right) + \left(\frac{1}{16} \times 8\right) + \left(\frac{1}{32} \times 10\right) + \cdots$$

The three dots are shorthand for "keep on adding more and more terms formed in the same manner: Keep doubling the denominators and increasing the multipliers by 2. Never stop."

Playing with Endless Sums

We now have two views of the average in which we are interested. First, we have data collected from volleyball games and tossed pennies. Second, we have an endless sum derived by common sense.

What does this seemingly endless sum mean? It makes no sense to add up an infinite number of numbers. No one can do that, not even with the aid of a computer executing billions of operations per second. It does make sense, however, to add up the first thousand terms in the sum, or the first million, or the first billion terms.

We can imagine adding up more and more terms and watching what happens to the sums. If they look as though they are getting closer and closer to a fixed number, we will call that number the *sum of the infinite number of terms.* As long as we keep in mind that we always add up only a finite number of terms, there should be no trouble with the misleading phrase "the sum of an infinite number of terms." The terms are said to form a *sequence,* or *series.*

With this warning in mind, let us tackle the "sum" of the series

$$\left(\frac{1}{2} \times 2\right) + \left(\frac{1}{4} \times 4\right) + \left(\frac{1}{8} \times 6\right) + \left(\frac{1}{16} \times 8\right) + \left(\frac{1}{32} \times 10\right) + \cdots$$

Let us add up the first six terms, just to get a sense of how the sums behave:

$$\left(\frac{1}{2} \times 2\right) + \left(\frac{1}{4} \times 4\right) + \left(\frac{1}{8} \times 6\right) + \left(\frac{1}{16} \times 8\right) + \left(\frac{1}{32} \times 10\right) + \left(\frac{1}{64} \times 12\right)$$

This sum is

$$\frac{2}{2} + \frac{4}{4} + \frac{6}{8} + \frac{8}{16} + \frac{10}{32} + \frac{12}{64}$$

or roughly

$$1 + 1 + 0.75 + 0.50 + 0.31 + 0.19$$

which is 3.75.

That is promising, for 3.75 is in the same ballpark as the experimental average for the pennies, 4.66. However, we still don't know what will happen as we add more and more terms. You may want to add a few more terms yourself, with or without the aid of a calculator.

Practice

Consider, for practice, a simpler sum in which the multipliers are all replaced by 1s—namely, the sum

$$\frac{1}{2} + \frac{1}{4} + \frac{1}{8} + \frac{1}{16} + \frac{1}{32} + \cdots$$

For instance, the sum of the first 4 terms, when each is expressed with a common denominator 16, is

$$\frac{8}{16} + \frac{4}{16} + \frac{2}{16} + \frac{1}{16}$$

which equals 15/16, or 0.9375. The sum of the first 10 terms is 0.9990 to four-decimal-place accuracy. It looks as though the sums are approaching 1.

A glance at the diagram to the right of a square cut into an infinite number of rectangles will show that the sum of the series is indeed 1. Unfortunately, only a few of the rectangles can be shown. The

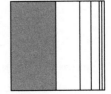

square has area 1. The shaded rectangle has area 1/2, leaving an area of 1/2 unshaded. The next rectangle fills half the unshaded area, namely, 1/4. That leaves an area of 1/4 that is not covered by those two rectangles. The next rectangle covers half that area, so has an area of 1/8 and leaves an area of 1/8 uncovered. The rectangles continue forever, each half as wide as the rectangle to its immediate left. The area that is covered by all the rectangles is 1, the area of the square, but it is also

$$\frac{1}{2} + \frac{1}{4} + \frac{1}{8} + \frac{1}{16} + \frac{1}{32} + \cdots$$

Hence the sum of this series is 1. To be precise, as we add more and more terms, we get sums that are closer and closer to 1.

From this fact we can obtain many more sums that will be useful in a moment. For instance, say that we start with the information that

$$\frac{1}{2} + \frac{1}{4} + \frac{1}{8} + \frac{1}{16} + \frac{1}{32} + \cdots = 1$$

Subtracting 1/2, the first term, from both sides tells us that

$$\frac{1}{4} + \frac{1}{8} + \frac{1}{16} + \frac{1}{32} + \cdots = \frac{1}{2}$$

Subtracting the first term, 1/4, from both sides of this equation tells us that

$$\frac{1}{8} + \frac{1}{16} + \frac{1}{32} + \cdots = \frac{1}{4}$$

A pattern is starting to show up: The sum, starting at any term, equals twice that term.

Finding the Average

Our detour through sums that are simpler than the sum

$$\left(\frac{1}{2}\times 2\right) + \left(\frac{1}{4}\times 4\right) + \left(\frac{1}{8}\times 6\right) + \left(\frac{1}{16}\times 8\right) + \left(\frac{1}{32}\times 10\right) + \cdots$$

will help us calculate this sum, which is the one that expresses the average number of tosses before heads or tails lead by 2.

First, let us divide each term by 2, in order to have a little simpler sum to deal with. Once we find the simpler sum, we must remember to multiply it by 2. The simpler sum is

$$\left(\frac{1}{2}\times 1\right) + \left(\frac{1}{4}\times 2\right) + \left(\frac{1}{8}\times 3\right) + \left(\frac{1}{16}\times 4\right)$$
$$+ \left(\frac{1}{32}\times 5\right) + \cdots$$

To find this sum, we will use a tool that will come in handy in Chapter 3, "The Complete Triangle." We will add up numbers in two different ways and then equate the two sums. The numbers will come from an infinite number of endless sums.

Using the information obtained earlier, in the practice section, we may write

$$\frac{1}{2}+\frac{1}{4}+\frac{1}{8}+\frac{1}{16}+\frac{1}{32}+\cdots = 1$$

$$\frac{1}{4}+\frac{1}{8}+\frac{1}{16}+\frac{1}{32}+\cdots = \frac{1}{2}$$

$$\frac{1}{8}+\frac{1}{16}+\frac{1}{32}+\cdots = \frac{1}{4}$$

$$\frac{1}{16}+\frac{1}{32}+\cdots = \frac{1}{8}$$

$$\cdots$$

Each horizontal line consists of a series of terms on the left of an equal sign and their sum on the right. The display shows the sums we get if we add the numbers left of the equal signs horizontally first.

This diagram displays the same idea pictorially.

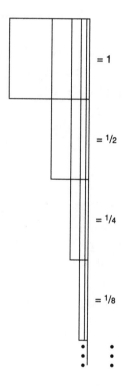

$$1 \times {}^{1}/_{2} + 2 \times {}^{1}/_{4} + 3 \times {}^{1}/_{8} + 4 \times {}^{1}/_{16} + \ldots$$

Now add up all the numbers to the left and right of the equal signs vertically first, column by column. That is, add them all by first adding up numbers that are directly above each other.

The first column has one 1/2. The second column has two 1/4s. The third column has three 1/8s. The pattern will go on through all the columns. Thus the sum of all the terms on the left is

$$\left(\frac{1}{2}\times 1\right) + \left(\frac{1}{4}\times 2\right) + \left(\frac{1}{8}\times 3\right) + \left(\frac{1}{16}\times 4\right) + \left(\frac{1}{32}\times 5\right) + \cdots$$

On the right we have the sum

$$1 + \frac{1}{2} + \frac{1}{4} + \frac{1}{8} + \frac{1}{16} + \frac{1}{32} + \cdots$$

The sum of all of its terms after the first is 1. So its sum is 1 + 1, or 2.

Since adding line by line will give the same sum as adding column by column, we find that

$$\left(\frac{1}{2}\times 1\right) + \left(\frac{1}{4}\times 2\right) + \left(\frac{1}{8}\times 3\right) + \left(\frac{1}{16}\times 4\right)$$
$$+ \left(\frac{1}{32}\times 5\right) + \cdots = 2$$

We double this (since we divided by 2 to simplify matters) to obtain the sum that equals the average number of tosses until there is a lead of 2. We finally have

$$\left(\frac{1}{2}\times 2\right) + \left(\frac{1}{4}\times 4\right) + \left(\frac{1}{8}\times 6\right) + \left(\frac{1}{16}\times 8\right)$$
$$+ \left(\frac{1}{32}\times 10\right) + \cdots = 4$$

In short, the theoretical average number is 4. The experimental average, based on 64 tosses, was 4.66. The two averages are not far apart. Presumably, if we toss a penny thousands of times, the observed average would gradually move closer to 4 and away from 4.66. In any case, our common-sense method yields numbers that are in reasonable agreement with those from the tossed pennies.

A Second Look at Volleyball

What has our common-sense approach told us about volley-ball games that are tied at 24 to 24? First, about half of them should end with a score of 26 to 24. Our record of 64 games included only 24 with that score, 8 less than theory suggests. Second, our theory predicts that one quarter of the games will end after 4 points are played, in other words, with a score of 27 to 25. This implies that three quarters of the 64 games, or 48, will end quickly, with at most 4 points scored after the 24-to-24 tie. Only 36 ended that soon. Third, the average number of points will be 4 even though some games will last quite long. That's far from the average of 7.03 in the 64 games.

The common-sense approach works fine for the tossed penny, but its volleyball predictions seem far off the mark. Why should this be? Why do so few volleyball games end as quickly as our analysis suggests? It seems that the volleyball games last longer than the pennies or our common-sense approach advise.

Volleyball enthusiasts offer a convincing explanation. In a rally, the receiving team has a much better chance than the serving team of scoring a point. The players know this, and if they win the opening coin toss, usually they choose to receive rather than serve.

Even though the teams are evenly matched, at any given rally the receiving team has an advantage, winning from 70 to 80 percent of the rallies in men's games and about 55 percent in women's games. Because of this, there will tend to be long stretches where teams alternate in scoring, thus taking longer to achieve a 2-point lead.

The mathematics that takes into account this contrast between server and receiver is more complicated than the approach we used, so we will not go into it. I will just state two of the conclusions that can be drawn from it. First, that the average number of points played after the tie is equal to "2 divided by the fraction of rallies the server wins." That fraction is in the range of 1/5 to 1/3. If neither the receiver nor the server has an advantage, then the fraction is 1/2, and we have the case examined in this chapter, represented by the flipped penny. Note that "2 divided by 1/2" is 4, in agreement with our calculations.

In some volleyball leagues a team must be serving to gain a point. If it is not serving and it wins a rally, it earns the right to serve but does not score a point. The mathematics that corresponds to these rules is much more complicated than our method. It turns out that if the teams are equally matched and neither server nor receiver has an advantage, then the serving team will score the next point in 2/3 of the cases, though many rallies may be played before it earns that point. Moreover, 2/3 of tied games will end as soon as possible, with only 2 points played. Two ninths will end with only 4 points played, and 2/27 with 6 points played.

I compared these theoretical numbers with the record of the national women's team. Of 54 tied games, 41 ended after 2 points, compared to a predicted 36; 10 ended after 4 points, compared to an expected 12; 3 ended after 6 points, compared to a theoretical 4. Clearly the theory gives a pretty good approximation of reality.

Other Views

The problem that we introduced through sports appears in other ways, such as in the study of wandering bugs or gambling. It is a part of probability theory and statistics called the *random walk,* which has been applied to epidemics, the stock market, and changes in the human population.

Imagine a bug who is free to travel between two parallel lines that are 4 units apart, as shown here. (Any animal would do, but, as in Chapter 1, bugs are most often called on to assist our intuition.)

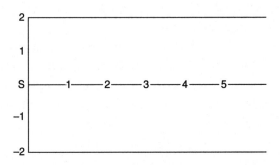

The bug starts at point *S* and then makes one of two permitted moves at random. In one move he goes 1 unit right and 1 unit up. In the other, he goes 1 unit right and 1 unit down. If he bumps into one of the two border lines, his trip is over.

His trip could be over in just two moves. This happens if he is unlucky and both moves are up or both are down. But if one is up and one down, he is back to the initial line and his trip continues.

Assume that the bug travels at random, as likely to move up as down. How many moves will he take, on average, before he bumps into a border line? We have already figured out the answer, which is 4 moves of the type described.

The same problem arises in the study of gambling, one of the earliest applications of the theory of probability some three centuries ago. Each of two gamblers has 2 chips. Before every play each gambler puts 1 chip in the pot. The players are equally likely to win and take the pot. What is the average number of plays before one gambler has all the chips? The answer, once again, is 4.

The theory has been worked out for situations in which the gamblers start with other, even differing, amounts, such as 3 chips and 5 chips. In this particular case the gambling will last on the average for 15 plays. More generally, the average number of plays is simply the product of the number of chips the two gamblers have at the start. I would hope that such a simple formula has a simple proof, but I have never heard of one.

The same problem appears when a gambler bets against the house. Say that he starts with 5 chips and wagers 1 chip at a time. Assume he has an equal chance of losing that chip or gaining 1 chip. He resolves to play until he is up to a total of 8 chips or is wiped out. On the average, if he faces this situation many times, he will play 15 times before his stake grows to 8 chips or else disappears.

A Remark on Endless Sums

In our study of volleyball or coin tossing, we showed that

$$\left(\frac{1}{2} \times 2\right) \ + \ \left(\frac{1}{4} \times 4\right) \ + \ \left(\frac{1}{8} \times 6\right) \ + \ \left(\frac{1}{16} \times 8\right)$$
$$+ \ \left(\frac{1}{32} \times 10\right) \ + \ \cdots = 4$$

This equation says that as we add more and more terms, the sum gets closer and closer to 4.

This suggests some questions not related to volleyball.

If we have an endless sequence of positive numbers that are getting closer and closer to zero, what can we say about their sums as we add more and more of them? Do these sums always tend toward some fixed number? If they do, can we always find that number, expressing it in terms of familiar numbers, as we did with the series in this chapter?

Or could the sums get arbitrarily large, eventually growing beyond any fixed number, even though the numbers we add get smaller and smaller?

For instance, what can we say about the following sequence?

$$\frac{1}{1} + \frac{1}{2} + \frac{1}{3} + \frac{1}{4} + \frac{1}{5} + \cdots$$

As we add more and more terms, the sums increase. One of two cases can occur. Perhaps the sums grow bigger and bigger, eventually exceeding 1000, then 1,000,000, and so on. After all, since we are adding positive numbers, this is a possibility.

On the other hand, the terms are getting smaller and smaller as we move along the series. Perhaps the sums don't grow arbitrarily large, but instead they are approaching some number, as was the case with the series that arose earlier in this chapter.

When I first met this series, whose terms are the reciprocals of all the whole numbers, I computed a few sums. On the basis of that information, I bet a fellow student that the sums stayed fairly small, never growing beyond 13. I lost the bet. A simple picture shows that the sums get arbitrarily large.

In the figure below each term in our series is represented by the area of a rectangle of appropriate height and width 1.

The area of this staircase is the sum of the series, whether infinite or finite.

We now draw another staircase made up of one 1-by-1/2 rectangle, two 1-by-1/4 rectangles, four 1-by-1/8 rectangles, eight 1-by-1/16 rectangles, and so on, as shown here:

Our second staircase fits inside the first one thus:

The total area of the shaded rectangles is

$$\frac{1}{2} + \frac{2}{4} + \frac{4}{8} + \frac{8}{16} + \frac{16}{32} + \cdots$$

which is

$$\frac{1}{2} + \frac{1}{2} + \frac{1}{2} + \frac{1}{2} + \frac{1}{2} + \cdots$$

Adding up 1/2s, which go on forever, we can make sums as large as we please. Thus the shaded area is infinite. It follows that the sum of the reciprocals of all the whole numbers, being even larger, is also infinite.

It isn't always easy to figure out whether a series of positive numbers will have an infinite sum or a finite sum. Entire books are devoted to techniques for deciding. Moreover, once we know a series has a finite sum, it isn't always easy to tell what that sum is, although we can calculate it to any number of decimal places we may choose.

For example, it has been known for over two centuries that the sum of the reciprocals of all the square numbers is one sixth of the square of π, the perimeter of a circle of diameter 1:

$$\frac{1}{1^2} + \frac{1}{2^2} + \frac{1}{3^2} + \frac{1}{4^2} + \frac{1}{5^2} + \cdots = \frac{\pi^2}{6}$$

The reasoning is quite ingenious. On the other hand, no one has a formula for the sum of the reciprocals of all the cubes of whole numbers:

$$\frac{1}{1^3} + \frac{1}{2^3} + \frac{1}{3^3} + \frac{1}{4^3} + \frac{1}{5^3} + \cdots$$

It was our good fortune to be able to find the exact sum of the series that gave us the average number of tosses of the penny. That is a rare piece of luck.

CHAPTER THREE

The Complete Triangle

It is hard to predict when or how a new invention or discovery will eventually be put to use. This chapter, which concerns the labeling of dots by letters, offers an example. Developed to prove theorems in topology, the result has since been applied in such varied topics as the fair division of an asset or cost among several people and the tiling of a polygon by triangles of equal areas. After we have developed the theory, we will describe some of these applications in more detail.

The Problem on a Line

Imagine a string of a's and b's that starts with an a and ends with a b. Here are two examples:

$$abaabbab \quad \text{and} \quad abbaaababbaab$$

The string can be short or long. What can be said about the number of times an a and a b are next to each other? This simple question begins a journey that takes us into a famous theorem called *Sperner's lemma.*

The first of our two strings has five such pairs of *a* and *b;* the second, seven. I invite you to conduct your own experiments. If your bookkeeping is accurate, you will find that the number of such pairs, where *a* and *b* are next to each other, is always odd. Why is this so? Before reading on, you might come up with your own reason.

Here is one explanation. I'll illustrate it with the first string, *abaabbab,* and begin by spreading it out as an *interval* on a line.

The interval is cut into seven shorter sections. Any section that has an *a* at one end and a *b* at the other end we call *complete.* A complete section occurs when we have an *ab* or a *ba* in our original string of *a*'s and *b*'s. So what we want to show is that the number of complete sections is odd. To do this, inside each section we place a pebble next to any of its ends labeled *a.*

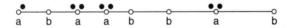

We shall count the total number of pebbles in two ways, just as we added up a collection of numbers in Chapter 2, "Win by Two." First we look at the pebbles from the viewpoint of the sections. In an *aa* section there are two pebbles; in a *bb* section there are none. A complete section has exactly one pebble. So, whether the total number of pebbles is odd or even depends only on the number of complete sections. If the number of complete sections is odd, so is the total number of pebbles. If the number of complete sections is even, so is the total number of pebbles.

Now we look at all the pebbles from the viewpoint of the dots, which are labeled *a* or *b*. There are no pebbles next to a *b*. Next to each interior *a* are two pebbles. Next to the dot labeled *a* at the end of the interval is one pebble. All told, the total number of pebbles is odd.

Putting the two counts together shows that the number of complete sections must be odd, so the number of times we have an *a* next to a *b* in the string must be odd. This means that behavior at just the two ends of the interval—an *a* at one end and a *b* at the other—influences behavior throughout the interval. If both ends had been labeled *a* or both had been labeled *b*, there would be an even number of complete sections. We summarize these observations by saying that if the number of *a*'s at the two ends is odd, so is the total number of complete sections. And if it's even, so is the total number of complete sections.

There is another way of looking at the strings, one that will give us more information. Now we will distinguish two types of complete sections, an *ab* section and a *ba* section. In an *ab*, the *a* is to the left of the *b;* in a *ba* the *a* is to the right of the *b*. We'll illustrate this approach with the same string, *abaabbab*, but describe it in general terms.

Let's start with the string stretched out on a line thus:

a	b	a	a	b	b	a	b

Starting at the left end, we walk along the interval until we meet a *b*. When that has happened, we have just crossed an *ab* section. Then we continue walking to the right until we meet an *a*. (If there are no more *a*'s, we're done: In such a case there is one *ab* section and no *ba* sections.)

However, when we meet another *a,* at that point we will have just walked across a *ba* section. Then we continue walking until we meet another *b.* And so on. On such a trip we run through all the complete sections, alternating *ab* sections and *ba* sections. Since the first and last sections we meet are *ab* sections, there is one more *ab* section than *ba* sections. Since the sum of an integer and the next higher integer is odd, we know that when we add the number of *ab* sections to the number of *ba* sections, the total number of complete sections will be odd.

The walking argument gives us the same conclusion as the pebble argument. However, it tells us a little more: There is one more *ab* section than *ba* sections. In a moment, as we move up from the interval to the plane, we will again have two ways of approaching a problem. And, again, one will tell us more than the other.

The Problem for Polygons

So far we have looked at subdivisions of an interval, a one-dimensional object. Now let's go up one dimension and consider subdivisions of a polygon by triangles. At this point we will need, in addition to *a* and *b,* the letter *c.* This is one of the few occasions when we choose to go so far into the alphabet, but, as you will soon see, doing so takes us into a fascinating world.

Consider a polygon, as below. In this chapter a *polygon* is a figure made up of straight pieces that form a circuit (or *loop*). The term also refers to the region that is bounded by that circuit.

Place a dot at each of the corners, and then place some more dots, at random, on the border and in the interior of this polygon. The figure that appears to the right illustrates just one of the countless ways that this can be done.

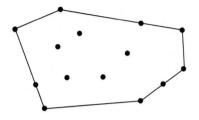

Then, using the dots as corner vertices, cut the polygon into nonoverlapping triangles. (This operation, too, can be done in many different ways, even for a given set of dots.)

Next, label each dot *a*, *b*, or *c* in any way whatsoever.

Depending on how their corners are labeled, there are 10 types of tri-

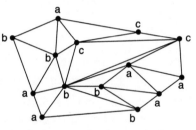

angles that may appear in the polygon, as is shown below.

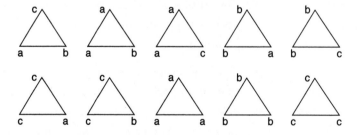

In only one of the 10 types do all three letters appear. Call such a triangle *complete*. Six types have two different letters, and three types have only one letter.

The complete triangle has exactly one edge with the letters *a* and *b*, a complete edge. (Recall that a complete edge is one whose ends are labeled *a* and *b*.) The triangle with two *a*'s and a *b* has two complete edges; so does the triangle with two *b*'s and an *a*. The others have no complete edges. We summarize this information below by placing a number inside each triangle to record the number of its complete edges.

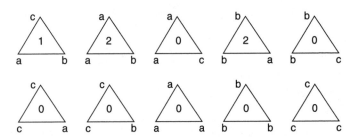

The complete triangle is the only one that has an odd number of complete edges—namely, 1. In a moment we will put this fact to use in another pebble-counting argument.

Our polygon happens to have two complete triangles and four complete edges on the boundary. I invite you to carry out a similar count with a polygon and labeling of your choice. If you do the counting accurately, your numbers should illustrate the following claim, which we will call the generalized *Sperner's lemma:*

> *If there is an odd number of complete edges on the boundary, then there is an odd number of complete triangles. If there is an even number of complete edges on the boundary, then there is an even number of complete triangles.*

In short, the number of complete boundary edges and the number of complete triangles are both odd or both even.

Emanuel Sperner (1905–1980) stated his lemma for triangles, the form he needed. In the 1928 paper where the lemma appeared, he used it to obtain short and elegant proofs for some deep topological theorems about the dimensions of various spaces. (The lemma can be generalized to all dimensions. For instance, in the next dimension up, dimension 3, his lemma refers to a solid tetrahedron, bounded by four triangles, and it involves the letters a, b, c, and d. In dimension 1, it concerns the interval, the case that opened the chapter.)

Proof

The pebble counting used for the interval, slightly modified, will show why the generalized Sperner's lemma is true. Instead of placing a pebble next to a dot labeled a, we place it next to a complete edge, labeled ab.

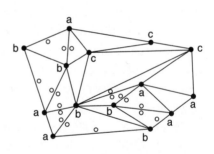

Inside each triangle in the subdivision of a polygon, we place a pebble next to each complete edge, as shown to the right. Each triangle will hold 0, 1, or 2 pebbles. Only the complete triangles have an odd number of pebbles. So, whether the total number of pebbles is odd or even is determined by whether the number of complete triangles is odd or even.

That counting was done from the viewpoint of the triangles. Next, count the pebbles by looking at the edges along which the pebbles lie. Along each complete edge on the border, there is one pebble. Along each complete edge in the interior of the polygon, there are two pebbles. These two

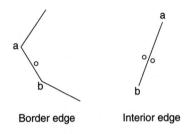

Border edge   Interior edge

types are shown to the left. Therefore the total number of pebbles is odd or even according to whether the number of complete boundary edges is odd or even. By comparing the two ways of counting—by triangles and by edges—we obtain the generalized Sperner's lemma.

Sperner's Special Case

Sperner, and many others since, used a very special case of his lemma, where the polygon is simply a triangle whose three corners are labeled *a*, *b*, and *c*. In this case, we introduce dots on the border and in the interior of the triangle and use them to cut the triangle into smaller triangles. Label dots that lie on the edge joining *a* and *b* either *a* or *b* at random. Dots on the edge joining *b* and *c* label either *b* or *c* at random. Label dots on the edge *ca* either *a* or *c* at random. Sperner's special lemma then asserts that *there must be at least one complete triangle among the smaller triangles that lie within the original triangle.*

We have already done the work to show that the lemma is true. First observe that *ab* sections on the boundary occur only on the edge that joins the corners labeled *a* and *b*. As we saw at the beginning of the chapter, there is an odd number of *ab* sections on that edge, for it is an interval in its own right with ends labeled *a* and *b*. The two other edges of the triangle have no *ab* sections. So the number of *ab* sections on the boundary of the triangle is odd. It follows immediately that the number of complete triangles is odd. That implies that there is at least one complete triangle, for zero is an even number.

In his 1928 paper, Sperner used this special case to establish a theorem about three subsets (parts) of a triangle whose vertices are labeled *a*, *b*, and *c*. Call the three parts $S(a)$, $S(b)$, and $S(c)$. These sets may be bounded by curves, or they could be much fancier, even resembling clouds of dust particles. Assume that $S(a)$ does not meet the edge *bc*, $S(b)$ does not meet the edge *ac*, and $S(c)$ does not meet the edge *ab*. Assume also that each point in the triangle is in at least one of the three parts. A point may belong to more than one set. An example of three such sets is shown here.

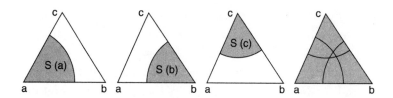

Suppose you are given a very small positive number *d*. Then Sperner's special lemma shows that *there are three points in the triangle, P, Q, and R, all within a distance* d *of each other, such that* P *is in* S(a), Q *is in* S(b), *and* R *is in* S(c).

The proof is quick, what is usually dubbed a "one liner," though such proofs usually take a few lines. Cut the given triangle into equilateral triangles whose sides have length less than *d*.

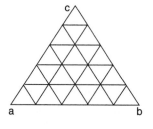

Place a dot at each vertex of each triangle. Label each dot *a*, *b*, or *c* according to which of the three sets it is in. For instance, if it is in $S(a)$, label it *a*. If it is in more than one of the sets, choose one of the possible labels. It is not hard to check that dots on the edge *ab* of the given triangle are labeled *a* or *b*.

Similar statements hold for the other two edges. Since the conditions of Sperner's lemma for a triangle are satisfied, there must be a triangle in the subdivision with all three labels. The corners of this triangle are within a distance d of each other and serve as the three promised points.

A Refinement

In the one-dimensional case, the interval with a at the left end and b at the right end, we saw that there is one more ab section than ba sections. That was a refinement of our first theorem, which referred only to complete sections and didn't distinguish the two types of complete sections. There is a similar refinement for our generalized Sperner's lemma.

First, there are two types of complete triangles. If we walk around the border of a complete triangle in the order a, b, and finally c, we move either in a clockwise or a counterclockwise direction. We may speak then of *clockwise complete triangles* and *counterclockwise complete triangles*, as shown here.

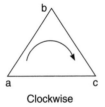
Clockwise

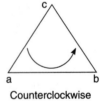
Counterclockwise

Now, imagine a polygon with dots labeled a, b, and c by which the polygon is cut into triangles. Then imagine walking around the border of the polygon in a counterclockwise direction. You may meet some complete edges. In some of these you will meet the a first and then the b. Call these ab edges. In some you will meet the b first and then the a. Call them ba edges.

The following turns out to be true:

The number of counterclockwise complete triangles minus the number of clockwise complete triangles equals the number of ab *edges minus the number of* ba *edges.*

(In our polygon on page 51 this checks out, for 1 − 1 = 2 − 2.) For instance, if there were, say, five *ab* edges and three *ba* edges, there would be two more counterclockwise complete triangles than clockwise. Therefore, there would have to be at least two complete triangles. Our original version of Sperner's lemma would not have told us that there has to be a complete triangle. It would only have told us that the number of complete triangles must be even since the number of complete edges (5 + 3) is even. But according to that version, we could not have ruled out zero, an even number, as the number of complete triangles. I'll mention later an application that requires the refined version of the lemma.

We can prove this refinement by slightly altering the pebble-counting technique. Instead of putting down pebbles, we put either a 1 or a −1 along each complete edge of each triangle. We place a 1 if, as we sweep out the triangle counterclockwise, we meet a complete edge in the order *a* first, then *b*, which we can call an *ab* edge. If we meet the complete edge as a *b* followed by an *a*, we place a −1, like this:

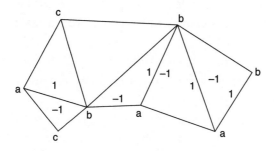

Now we sum all the −1s and 1s. We know that each counter-clockwise complete triangle contributes a 1; and each clock-wise one has a −1. Each of the other types contributes a total of 0. Therefore, the total of all the 1s and −1s is

the number of counterclockwise complete triangles minus the number of clockwise complete triangles.

Then we add up all the 1s and −1s from the viewpoint of the complete edges. Since an inner complete edge has a 1 and a −1, this sum is

the number of ab *edges on the border minus the number of* ba *edges on the border.*

That shows why the refinement holds [since the two ways of adding all the 1s and −1s must give the same sum].

A Constructive Proof

If there is an odd number of complete border edges, there must be at least one complete triangle. The proof we gave for this is what is called an *existence proof*. It shows that some-thing exists but doesn't tell how to find it. Often, for practi-cal computation, an existence proof is only tantalizing. What one may want is a *constructive proof,* one that gives a proce-dure for finding the object that we know exists. Of course, this then raises more questions, such as, "How efficient is the procedure?" "Is there a faster procedure?" "How fast can the fastest possible procedure be?" These are questions that a computer scientist faces.

But let's get back to describing a procedure for finding a complete triangle if there is an odd number of complete edges on the border.

We pick a complete border edge and start a journey, which will take us from triangle to triangle, at the triangle next to that edge. If that triangle is complete, stop. We have found a complete triangle. If that triangle is not complete, its third vertex is labeled either *a* or *b,* and the triangle has another complete edge:

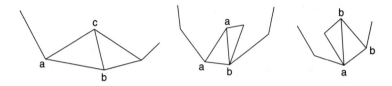

When the triangle is not complete, we cross the second complete edge into the triangle on the other side of that edge. If that adjacent triangle is complete, we stop; our search is successful. If it is not complete, we continue the journey following the same rule. We continue in this fashion, moving triangle by triangle. Our journey will stop either when we enter a complete triangle or arrive at a complete edge on the border, as illustrated here.

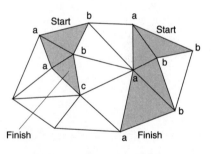

If we do not bump into a complete triangle, we repeat the procedure starting at another complete boundary edge. (Since there is an odd number of complete boundary edges, there will still be unused ones.) We continue sweeping out such paths. Each path that does not end at a complete triangle uses up two complete edges on the boundary: one at the start and one at the finish. Since there is only a finite number of triangles, we will eventually have a path that meets a complete triangle. (Not all the paths can end at the border, for then there would be an even number of complete edges on the border.) By the way, if there were only one complete edge on the border, our very first path would find the complete triangle.

Note that this process not only provides a way of finding the complete triangle, it also provides a second proof that such a triangle exists.

Two Detours

Let me add two detours, ones that use the *counting-two-ways technique.*

If we draw a polygon and some dots, we can use the dots in many different ways to cut the polygon into triangles. Below is an example in which a polygon is subdivided in two ways with the same dots.

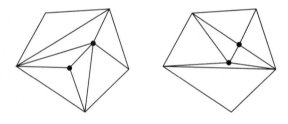

In each of these polygons there are seven triangles. This illustrates a general phenomenon, as you may check with a polygon and dots of your choice: *The number of triangles will be the same, no matter how you use a given set of dots to cut up the polygon.*

To see why, imagine any subdivision of a polygon based on a choice of dots. Each triangle in the subdivision contributes three angles that total 180 degrees. (A proof of this is in Appendix A.) Therefore the total number of degrees of all the angles in the subdivision is the product of the number of triangles and 180.

Now look at the angles from the point of view of the dots. At an inside dot the total of the angles meeting that dot is 360 degrees since they fill out a whole circle around that dot. At a dot on the border but not at a corner of the polygon, the angles total 180 degrees. The sum of the angles at the corner dots is completely determined by the polygon. (We don't need to know the value of this sum, just that it is the same in both cases.)

Since the total of all the angles is determined by the number of inside dots, the number of border dots, and the polygon, the total is the same for all subdivisions drawn with the aid of a given choice of dots. Since that total is also equal to the number of triangles times 180, the number of triangles is determined by the dots and the polygon. (Appendix A develops a formula for the number of triangles in terms of the numbers of inside dots, border dots, and corner dots.)

My second detour concerns cutting the surface of a sphere into curvilinear triangles. Imagine placing dots on the surface of a sphere and then using them to cover the sphere with triangles. *It turns out that there must be an even number of triangles.* I invite you to show this by counting pebbles in two ways. (Place three pebbles in each triangle, one along each edge.) Similarly, if you then label all those dots *a*, *b*, or *c* at random, there must be an even number of complete triangles. Pebbles will show this too. Or you could show it by taking a trip across *ab* edges as we did earlier.

Applications

It is not known how Sperner happened to think of his lemma. But once discovered, it became a tool that anyone could call on when needed. When mathematicians see their discoveries put to uses they had never imagined, it cheers them and reinforces their faith that their work is of value. I'm sure Sperner, when asked to participate in a 1979 conference, *Numerical Solutions of Highly Nonlinear Problems*, was delighted to see his lemma playing a role in applied mathematics.

Over the years Sperner's lemma has been applied in a surprising variety of problems. In 1929, soon after the lemma was published, three mathematicians, Knaster, Kuratowski, and Mazurkiewicz, used it to obtain a very short proof of a topological result known as *Brouwer's fixed-point theorem*, which went back to 1910. Roughly speaking, that theorem says that if you have a flat piece of rubber in the shape, say, of a square, and crumple it and stretch it and put it back within the area where it had been, then some point, called a "fixed point," is back where it started. For instance, if you just rotate the square 90 degrees about its center, the center is the sole fixed point. Brouwer's theorem has been applied in differential equations (part of advanced calculus) and the theory of matrices (part of advanced algebra).

The first application of Sperner's lemma, generalized, of which I am aware was published in 1970. Monsky proved that it is impossible to cut a square into an odd number of triangles of equal areas. (It is not hard to cut it into any even number of such triangles.) In 1989 Kasimatis proved that if n is an integer greater than 4 and you cut a regular n-sided polygon into triangles of equal areas, then the number of triangles must be a multiple of n. (When n is even, the refined version of Sperner's lemma is needed.)

In 1987 Sperner's lemma was the basis of a proof by Schmerl, one of 14 different proofs of the following geometric theorem: Assume that a rectangle is tiled by rectangles, each of which has at least one side whose length is a whole number. Then the tiled rectangle itself has at least one side whose length is a whole number.

In 1995 Hochberg, McDiarmid, and Saks used Sperner's lemma when studying a certain type of labeling of the dots in a system of dots and edges. I'll illustrate their work by one special case.

Consider the diagram of dots, edges, and triangles to the right. There are 15 dots.

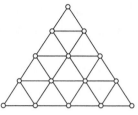

Label each of the dots with one of the numbers from 1 to 15, using each number once. This can be done in many ways. One way is the straightforward, orderly way shown to the right.

Each edge has a number at each of its two end dots. The largest difference between these two numbers is 5, which occurs several times; for instance, at the edge whose ends have an 8 and a

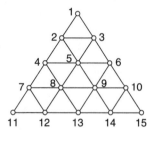

13. Is it possible to number the dots in such a way that all the differences are at most 4? The answer is no, and Sperner's lemma plays a role in showing this.

In 1999 Su published "Rental Harmony: Sperner's Lemma in Fair Division." It begins: "My friend's dilemma was a practical question that mathematics could answer, both elegantly and constructively. He and his housemates were moving to a house with rooms of various sizes and features, and were having trouble deciding who should get which room and for what part of the total rent. He asked, 'Do you think there's always a way to partition the rent so that each person will prefer a different room?' "

This type of problem has a long history. It occurs wherever several people want to divide some benefit or chore fairly. As Su remarks: "Our fair division approach is based on a simple combinatorial lemma, due to Sperner. However, do not be fooled—this little lemma is as powerful as it is simple."

When he came up with his lemma, Sperner was not thinking about fair division, fixed-point theorems, cutting a square into triangles or a rectangle into rectangles, or differential equations. Like an explorer of some uncharted land or sea, he added to our knowledge, moving us another step away from all other species, expanding the tool chest and the options of our civilization. That is typical of scientific discovery, whether it grows out of curiosity or out of necessity. And no one can predict how the many discoveries still being made today will be used in the unforeseeable future.

CHAPTER FOUR

Slumps and Streaks

In 1993 the Los Angeles Dodgers won exactly as many base-ball games as they lost. During a season of 162 games, they won 81 and lost 81. In this they resembled a penny tossed 162 times. But how close is this resemblance? I wondered whether the way the wins and losses occurred also resembles the way a tossed penny behaves. For instance, do the runs of consecutive losses or consecutive wins resemble the runs of heads or of tails one gets tossing a penny repeatedly? In a nutshell, how random were the Dodgers' wins and losses?

To find out, I contacted the team's publicity office, which quickly faxed me that year's record of wins and losses, game by game. These are the data, where *W* stands for a win and *L* for a loss:

LWWLLWLLLLWWWLLLLLLWWLLWLWWLWWLLWLLL
WWWWWWWWWWWLWWLWWLLWLLWWWWLLLWWLLLL
WWWLWWLWLWLWLWLWWWLWLLLLWWLWLWLWLWLL
WWWLLLLLLLWWLLWWWWWWLLLWLLWWWLWLL
WLLWLWWWLWLWWWLWLLLWLLLLW

As you scan the record, you will see isolated *L*'s and *W*'s along with much longer blocks, even a block of 11 *W*'s and a block of 7 *L*'s. A block of *W*'s we will call a *streak,* even if it has just one *W.* For a weak team such a block may indeed be viewed as a winning streak. A block of *L*'s of any length is a *slump.* There are 40 slumps alternating with 40 streaks.

To obtain an orderly overall view, we record the data in terms of the number of slumps and streaks of various lengths. First, since slumps and streaks alternate, and the first block is a slump and the last is a streak, the number of slumps must equal the number of streaks. That observation served as a check on my bookkeeping:

Length of Slump	Number of Slumps	Length of Streak	Number of Streaks
1	20	1	20
2	10	2	11
3	4	3	6
4	4	4	1
5	0	5	0
6	1	6	1
7	1	7	0
		8	0
		9	0
		10	0
		11	1

The most common slump or streak is just 1 game long. Half the slumps and half the streaks are of that length. The slumps or streaks of length 2 are about half as frequent. Together with those of length 1, they account for about three fourths of the blocks. Long blocks are rare, with only four of them containing more than 4 games.

Next let us lump the slumps and streaks together. The next table combines the previous two tables:

Length of Block	Number of Blocks
1	40
2	21
3	10
4	5
5	0
6	2
7	1
8	0
9	0
10	0
11	1

Out of the 80 blocks, exactly half have length 1, the shortest possible. About half as frequent are the 21 blocks of length 2. Half as frequent again are the 10 blocks of length 3. Blocks of length 4 are half as frequent as blocks of length 3. The few longer blocks are of fairly random lengths.

Recall that I picked the Dodgers of 1993 because they won as many games as they lost. If we flipped a penny 162 times, we would expect around 81 heads and 81 tails. There would be runs of heads and runs of tails. How closely would their frequencies resemble those we just tabulated for the Dodgers? For instance, would there be about 80 runs? Would about half of them have length 1? Would the long runs be rare?

An Experiment

There is a simple way to find out. Toss a penny 162 times and see what happens. Here are the results when I did just that.

There were 77 heads and 85 tails:

Length of Block of Heads	Number of Blocks	Length of Block of Tails	Number of Blocks
1	22	1	22
2	11	2	5
3	5	3	8
4	2	4	3
5	0	5	2
6	0	6	0
7	0	7	1
8	0		
9	0		
10	1		

Except for the large number of tails blocks of length 3, the data resemble the tables for the Dodgers. Combining the data for tails and heads, as we did for wins and losses, yields the following information about the blocks:

Length of Block	Number of Blocks
1	44
2	16
3	13
4	5
5	2
6	0
7	1
8	0
9	0
10	1

There are 82 blocks for the penny, a bit more than the 80 blocks for the Dodgers. In both cases the most frequent length is 1, and there are very few long blocks.

Let us compare the average block length for the two cases. The Dodgers had 80 blocks. Since there were 162 games distributed in these blocks, the average length of a block is 162/80, which is about 2.025. For the penny the average length is 162/82 since it had 82 blocks. That is about 1.976.

These comparisons suggest that chance played a big role in whether the Dodgers won or lost a game, that chance alone can generate long blocks, hence long slumps and streaks. In other words, slumps and streaks are to be expected, even though the skill of batters and fielders may not change. That chance plays a role should come as no surprise. For instance, the difference between a home run and a long fly caught at the fence may be just a few inches or a random puff of air.

A Common-Sense Approach

Experiments, whether with the aid of a baseball team or a penny, can only suggest conjectures. (A conjecture has more weight than a mere guess, for it is based on substantial evidence.) With the data we have collected, we may make some predictions about what will happen if a hypothetical penny is tossed billions of times. Mathematics can deal most easily with a rigid structure, such as arithmetic, or with the totally random. Oddly, within absolutely chance events a strict order can be discovered. However, a mix of the controlled and the chaotic presents a specially difficult challenge, for we then face the problem of extracting the signal from the noise. It is the same as the challenge faced by a doctor diagnosing a patient or an engineer getting rid of background hiss in a recording. Luckily, in the case of our hypothetical penny, we will deal with chance alone.

On the basis of the data at our disposal, we can make a few predictions about how our penny should behave. First, about half the blocks should have length 1. About a quarter should have length 2. Long blocks should be rare. The average length of a block should be around 2. If the theory we develop gives results much different from these, we would either have to carry out more experiments or alter the theory.

It may turn out that chance alone could cause long streaks and slumps. If so, then baseball players need not take them so personally, feeling depressed in a slump or ecstatic in a streak. Similarly, a gambler enjoying a run of good luck should not assume that someone above is taking an interest.

Imagine now an ideal penny, equally likely to land tails as heads. That means that if this penny is thrown trillions of times, it will come up heads on very close to half the tosses and tails on half the tosses. To distinguish this penny from the real penny that I tossed 162 times, I will denote heads by a and tails by b.

Our first question is, "What fraction of the blocks will have length 1?" Or, more specifically, "If the ideal penny is flipped trillions of times, about what fraction of the blocks will have length 1?"

Rather than examining blocks that appear later in the imaginary experiment, we consider only the first block, for it is typical of all the blocks. A block is a block, wherever it may occur. We are now asking, "What are the odds that the first block has length 1?"

Look closely at how a block of length 1 happens. The first toss of the penny can show either tails or heads, but the second toss must be the opposite of that first toss. Since the ideal penny is fair, the odds of being opposite is exactly one half. So we would expect half of the blocks to have length 1. This is in agreement with what we observed: For the

Dodgers, 40 of the 80 blocks had length 1; for the penny, 44 out of 82. So far, so good.

To strengthen our insight, let us analyze that first block another way. In this second method, we list all the ways the first 2 tosses could turn up, and see which ones lead to a block of length 1.

There are four equally likely cases:

aa, ab, ba, bb

In the second and third cases there is a head and a tail. That is, the second toss is the opposite of the first one, and a block of length 1 forms. There are therefore two out of four chances that the first block has length 1. Therefore, the odds that the opening block has length 1 is 2/4 or 1/2, in agreement with our first analysis.

What fraction of the blocks will have length 2? Again we focus on the opening block. This time we must consider the first 3 tosses. There are eight possible ways they appear:

aaa, aab, aba, abb, baa, bab, bba, bbb

In the first and last cases, *aaa* and *bbb*, the opening block will have length of at least 3. In the four middle cases, *aba, abb, baa,* and *bab,* the opening block has length 1. Only the cases *aab* and *bba* produce an opening block of length 2, that is, two cases out of eight equally likely cases. So the odds of having a block of length 2 are 2/8 or 1/4.

We could reach the same conclusion a little more directly. To form a block of length 2, two things must happen. The first 2 tosses must be the same, and the third must be different from them. In half the cases the first condition will hold. In half of those cases, in which the first condition holds, the second restriction also holds. The odds of both conditions being fulfilled are a half of a half, or 1/4.

Next, what are the odds that a block has length 3? Again there are 2 ways to answer. Using the long method, we could make a list of all 16 ways that the first 4 tosses can turn up. (I urge you to do this.) Then see in how many of these cases a block of length 3 forms. These are the cases *aaab* and *bbba*. That will give odds of 2 out of 16, that is, 2/16 or 1/8.

Or we could take a shortcut, as follows. To form a block of length 3, three things must happen: The second and third tosses must be the same as the first, and the fourth must be opposite. The odds for each of these things occurring is 1/2. Therefore, the likelihood that all three conditions are met is $1/2 \times 1/2 \times 1/2$, which, as we would expect, is again 1/8.

In the case of the Dodgers, 10 out of 80 blocks had length 3, which is exactly 1/8, or 12.5 percent. For the flipped penny, it was 13 out of 82, or about 15.9 percent, slightly larger than the theoretical 1/8.

In the same way we can show that the odds that a block has length 4 are $1/2 \times 1/2 \times 1/2 \times 1/2$, which is $(1/2)^4$ or 1/16. More generally, the odds that a block has length n are just the product of n of the 1/2s. No wonder a long slump or streak is rare—but they can occur.

The Average Block

The average block length for the Dodgers was 2.025 and for the tossed penny, 1.976. What average would common sense predict? We will figure out this theoretical average in two different ways.

Recall how the average was calculated for the Dodgers. It was the fraction 162/80, where 80 is the number of blocks and 162 the number of games. That number 162 is the sum of the lengths of all the blocks. Remember that there are 40 blocks of length 1, 21 blocks of length 2, and so on. Rewriting the numerator, 162, of the fraction 162/80 in terms of these various block lengths, we conclude that the average block length is

$$\frac{\begin{array}{c}(40 \times 1) \;+\; (21 \times 2) \;+\; (10 \times 3) \;+\; (5 \times 4) \;+\; (0 \times 5) \\ +\; (2 \times 6) \;+\; (1 \times 7) \;+\; (0 \times 8) \;+\; (0 \times 9) \;+\; (1 \times 10)\end{array}}{80}$$

Rewrite this as the sum of 10 fractions:

$$\left(\frac{40}{80} \times 1\right) + \left(\frac{21}{80} \times 2\right) + \left(\frac{10}{80} \times 3\right) + \left(\frac{5}{80} \times 4\right) + \left(\frac{0}{80} \times 5\right)$$
$$+ \left(\frac{2}{80} \times 6\right) + \left(\frac{1}{80} \times 7\right) + \left(\frac{0}{80} \times 8\right) + \left(\frac{0}{80} \times 9\right) + \left(\frac{1}{80} \times 10\right)$$

The first fraction, 40/80, is the fraction of blocks that have length 1. The next number, 21/80, is the fraction of blocks of length 2, and so on. To calculate the theoretical average—without any experiments—we replace these fractions by the theoretical frequencies we have just computed. The *theoretical average* is then

$$\frac{1}{2} \times 1 \;+\; \frac{1}{4} \times 2 \;+\; \frac{1}{8} \times 3 \;+\; \frac{1}{16} \times 4 \;+\; \frac{1}{32} \times 5 \;+\; \cdots$$

We carried out the same steps when analyzing volleyball games in Chapter 2.

We now face an endless sum, which looks a lot like a sum met in that chapter. To be precise, each term in this sum is exactly half the corresponding term in the sum already evaluated. Since that sum was 4, the new sum is 2. In short, the theoretical average is 2.

Both experimental averages of block lengths, 2.025 and 1.976, are quite close to the theoretical average. This agreement reassures us that our analysis makes sense.

There is an even shorter way to compute the theoretical average. We imagine a huge experiment in which the hypothetical penny is tossed, say, a million times. How many blocks would we expect?

As the penny is tossed, a new block starts whenever the penny comes up the opposite of what it did on the preceding toss. The odds of this happening are 1/2. That means about 500,000 times in a million tosses. So there should be about 500,000 blocks. The total of all their lengths is 1,000,000. The average length of a block is consequently about

$$\frac{1,000,000}{500,000} = 2$$

This is the same value that we obtained by summing an endless series.

We have two completely different ways of finding the average: by the shortcut and by the series. From this we again conclude that

$$\left(\frac{1}{2} \times 1\right) + \left(\frac{1}{4} \times 2\right) + \left(\frac{1}{8} \times 3\right) + \left(\frac{1}{16} \times 4\right) + \left(\frac{1}{32} \times 5\right) + \cdots = 2$$

Of course we already knew this, but we now have a short, indirect way of obtaining the same result. By considering the average of certain random events, we managed to sum an endless series.

Just as we studied the slumps and streaks of a team, we could analyze them for an individual batter. The mathematics is more complicated because the chance of getting a hit does not equal the chance of getting an out. (According to the *Baseball Encyclopedia*, the typical batting average in the major leagues is about 0.255: a player will get a hit in about 1 in 4 at bats.)

In the case of basketball this analysis has already been done. The results were reported in 1985 in the journal *Cognitive Psychology* in an article titled "The Hot Hand in Basketball: On the Misperception of Random Sequences" by Gilovich, Vallone, and Tversky. Based on the performance of the Philadelphia 76ers and the Cornell University varsity teams, they concluded that the slumps and streaks of individual players were close to what would be predicted on the basis of chance alone. They advised basketball coaches, who believe in the "hot hand":

> Passing the ball to the player who is "hot" is a common strategy. It is also anticipated by the opposing team who can concentrate on guarding the "hot" player. If another player, who is less "hot" on that particular day, is equally skilled, then the less guarded player would have a better chance of scoring. Thus the belief in the "hot hand" is not just erroneous, it could also be costly.

Endless series, under their official name *infinite series*, are common in the study of chance events and in such varied fields as economics, trigonometry, and physics. The applications in this and the volleyball chapter (2) are only a small sample of their many uses.

CHAPTER FIVE

Thrifty Strings

To measure the distance to Venus, scientists bounced a radar signal off its surface and clocked how long it took the signal to return. They knew that if they used a single pulse, it would get lost in the prevailing static. Therefore, the scientists sent a string of pulses in a pattern that they could recognize even if part of the string became garbled. This chapter develops the mathematics behind that pattern. At the end of the chapter I will describe other applications, some of which we use daily without being aware of them.

We begin again with a simple question about strings of *a*'s and *b*'s.

The Question

Before we can pose the question, we have to define some terms. Any group of consecutive symbols in a string we call a *word*. The *length* of a word is the number of symbols in it.

A word of length 2 we call a *couplet,* or *2-tuplet.* A word of length 3 is a *triplet,* or *3-tuplet.* A word of length 4, a *quadruplet,* is then a *4-tuplet,* and a word of length 5, a *quintuplet,* is a *5-tuplet.* More generally, for any whole number $n = 2, 3, 4, \ldots$, a word of length n we call an *n-tuplet.*

The string

aababbaaaba

is itself a word of length 11. It contains 11 words of length 1. It also contains 10 (overlapping) couplets, 9 triplets, 8 4-tuplets, and so on. Thus it has words of all lengths from 1 to 11.

We start with the simplest version of the general question:

How long can a string be if no couplet in it appears more than once?

Let's see how long a string we can build by adding one letter at a time. We might as well start with the letter *a.* Because we are only on the first couplet, the next letter can be either *a* or *b.* Let's use *a* again, producing *aa.* The next letter cannot be *a* because then our string would become *aaa* where the couplet *aa* appears more than once. It is safe, though, to add *b* next, and obtain the string *aab.* This string contains the couplets *aa* and *ab.*

For the fourth letter we can add either *a* or *b* without repeating any couplet. Let's use *a,* getting *aaba.* The fifth letter cannot be *a* since then the couplet *aa* would be repeated. Nor can it be a *b,* for then the couplet *ab* would appear twice. We should backtrack and change the fourth letter to *b* and see what happens. We then have *aabb.* We can't add *b* because then the couplet *bb* would appear twice. Adding *a* is permissible, though, and doing so gives us the string *aabba* of length 5, with no repeated couplet. You may experiment yourself to build other strings of length 5 with no repeated couplet.

If you try to construct a string of length 6 with no repeated couplet, you will be disappointed. To see why, note that there are only four possible couplets of *a* and *b*. They are *aa, ab, ba,* and *bb*. In any string of length 6, each of the first five symbols in the string would be the first symbol of a couplet. That forces the string to contain five couplets. Because there are only four possible couplets of *a* and *b*, this means at least one couplet must appear more than once as soon as the string has five couplets (or six symbols).

Now that we have considered couplets, the next question that invites an answer is,

How long can a string be that has no duplicated triplet?

To find out, we experiment again, starting, say, with the string *abab*. The next letter cannot be *a*, because the string *ababa* would contain the triplet *aba* twice. Choosing *b* gives no trouble, so we use it and reach *ababb*.

Next we can add either *a* or *b*. Choosing *b* gives us the string *ababbb*. If we add *b* next, we will have the string *ababbbb* in which the triplet *bbb* appears twice. But we can use *a*, obtaining the string *ababbba*. Putting *b* next would force a repeat of *bab*, but we can use *a*, and we have the string *ababbbaa* of length 8. The next symbol can be either *a* or *b*.

If we choose *b*, we get *ababbbaab*. But then we are stuck, for if we add *a*, the triplet *aba* repeats, and if we add *b*, the triplet *abb* repeats. So we back up and use an *a* instead of *b*, giving us a string of length 9:

ababbbaaa

If we add *a* to this string, then we repeat the triplet *aaa*. But we can safely add *b*, obtaining a string of length 10 without repeated triplets:

ababbbaaab

It turns out that any string of length 11 (or greater) will inevitably contain a repeated triplet. The explanation for this is just a slight variation of the one we just gave for strings that have no repeated couplet:

Imagine a string of length 11, composed of *a*'s and *b*'s, occupying the 11 places here:

_ _ _ _ _ _ _ _ _ _ _

Each of the first 9 symbols is the first symbol of a triplet in the string. The final 2 symbols merely complete the last triplet. If there were no duplicated triplet in an 11-symbol string, there would need to be at least 9 different triplets made up of *a*'s and *b*'s.

Let's make a list of all possible triplets of *a* and *b* to see how many exist. We arrange them in dictionary order to be sure we find them all:

aaa	*baa*
aab	*bab*
aba	*bba*
abb	*bbb*

All told, in this "dictionary list" we found 8: 4 beginning with *a* and 4 beginning with *b*. In short, we can create all possible triplets by putting either *a* or *b* in front of each of the 4 couplets we identified earlier. That means that the number of triplets is twice the number of couplets—hence 2 times 4, or 8. We really didn't need to make the list; we could just have multiplied by 2 the number of possible couplets.

Since any string of length 11 would contain 9 triplets, at least 1 triplet in it would have to be repeated. So the longest a string without repeated triplets can be is 10.

We have found a string that contains each triplet exactly once and a string that contains each couplet exactly once. These are examples of what we will call *full strings*. For a given whole number *n*, a full string is a string that contains each *n*-tuplet exactly once.

We have found that there are full strings when *n* is 2 or 3. Is there a full string for 4-tuplets? For 5-tuplets? If so, how many symbols would there be in them? I invite you to experiment with pencil, paper, and eraser.

From Strings to Wheels

Look at the full string for triplets that we have just found:

ababbbaaab

Note that the final couplet *ab* is the same as the opening couplet *ab*. This is not a coincidence. It will be true for any full string for triplets. Here is why: Imagine any full string for triplets, ending, say, in *ab*, like this:

_ _ _ _ _ _ _ _ *a b*

We will show that the first couplet has to be *ab* also. In other words, we will show that the first triplet must always be either *aba* or *abb*.

Our approach will be indirect. We will assume that neither of those two triplets appears as the first triplet in the full string. From this we will show that there would then be at least three different triplets that end in the couplet *ab*. Such a conclusion is nonsense because there are only two triplets that end in *ab*.

If neither of these triplets, *aba* and *abb*, appears at the beginning of the string, they must both appear later, somewhere in _ _ _ _ _ _ _ _ *a b*. Then each has an immediate predecessor, overlapping it in the couplet *ab*. One is *aab*, and the other is *bab*. That tells us that the triplets *aab* and *bab* are not the final triplet in the string. But the final triplet also ends in *ab*. That would mean that there are at least three triplets that end in *ab*. But we know that there are only two triplets that end in *ab*—namely, *aab* and *bab*. So our assumption that the opening couplet is not *ab* has produced a false statement. That forces us to conclude that the opening couplet must be *ab*, the same as the final couplet.

The same reasoning shows that in any full string for triplets, the first and last couplets will always be the same. To summarize, let *pq* stand for any one of the four possible couplets, *aa*, *ab*, *ba*, and *bb*. That means that *p* stands for either *a* or *b*, and that *q* stands for either *a* or *b*. Using the letters *p* and *q* as substitutes for either *a* or *b* permits us to make general statements economically.

We see that any full string for triplets begins and ends with *pq*, as shown below. Since *p* can be either *a* or *b*, and *q* can be either *a* or *b*, this string is short for four different possibilities:

$$p\,q_____p\,q$$

Because the couplets at the ends of this string are the same, we can bend the string into a circle. The figure below shows

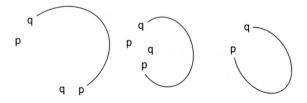

how this is done, first gradually forming a spiral, then figuratively gluing the matching ends *pq* together so the two *p*'s merge and so do the two *q*'s.

We can change any full string for triplets into a wheel with eight symbols in this manner. For instance, the full string for triplets, *ababbbaaab*, becomes the wheel shown here. (Read the wheel clockwise, starting at the leftmost *a*.)

```
          a
  b              b

a                    b

  a              b
          a
```

In this wheel you can start anywhere you want and run through all eight triplets one by one. With a mere 8 symbols the wheel lists all eight triplets. This is fewer than the 10 symbols of a full string and far fewer than the 24 (8 times 3) symbols that appear in our dictionary list of triplets.

We now have three ways to list the eight triplets of *a* and *b*: a dictionary list, a full string, or a wheel. Not only is the wheel the most economical way to proceed, but it also turns out to be the most convenient to analyze, as we will soon see.

A Closer Look

Before we go on to examine strings and wheels for quadruplets, it would be prudent to take a closer look at what we have done with triplets. The fresh perspective we obtain will give us a new tool for dealing with *n*-tuplets for any value of *n*.

What makes a full string so efficient is that it exploits overlaps. For instance, the first four symbols, *abab*, of our full string, *ababbbaaab*, records the triplets *aba* and *bab*, which overlap in the couplet *ba*. We can see all the overlaps of adjacent triplets in our full string by spreading the triplets out in a way that clearly displays how each triplet overlaps most of the next triplet:

aba

bab

abb

bbb

bba

baa

aaa

aab

To describe this overlapping in detail, we introduce a useful notation. As we did a moment ago, we will use letters other than *a* and *b* to stand for either of those two letters. We can then call the general triplet *pqr*. An overlapping triplet following it would share the couplet *qr*. So it would have the form *qrs*. Here each of *p*, *q*, *r*, and *s* stands for either *a* or *b*. With this in mind, let's turn our attention to strings for 4-tuplets (quadruplets).

To begin, we count the number of 4-tuplets by using the same method with which we counted the number of riplets. We can obtain all quadruplets by putting an *a* or *b* in front of each of the 8 triplets. That tells us that there are twice as many quadruplets as triplets, hence 16 of them.

How long would a full string for quadruplets need to be? Each of the first 16 symbols in such a string would start a quadruplet, and the final 3 symbols would complete the last quadruplet. The string would therefore have a length of 16 + 3. In other words, it would contain 19 symbols.

Just as we showed that the final couplet in a full string for triplets is the same as the opening couplet, we could show that the final triplet in a full string for quadruplets would always have to be just the same as the opening triplet. This means that the full string could be bent into a wheel with 16 symbols. Before reading on, you may want to experiment by trying to make such a wheel. Even if you are not successful, the practice will help you understand the following discussion.

Rather than just struggle until we manage to build such a wheel, we will take our time and analyze what is going on in such a wheel. The insight we gain will help us treat wheels for much longer n-tuplets, without any further thought or work.

Imagine that in such a wheel the word

abbabbb

occurs. This part records the following four 4-tuplets:

abba, bbab, babb, abbb

The first two of these 4-tuplets overlap in the triplet *bba*. We can record this overlap in a diagram, like this:

$$\text{abba} \longrightarrow \underset{\text{bba}}{\circ} \quad \text{bbab} \longrightarrow$$

Next, let's look at the quadruplets *bbab* and *babb*. They overlap in the triplet *bab*. We extend the diagram to include this information:

$$\text{abba} \longrightarrow \underset{\text{bba}}{\circ} \xrightarrow{\text{bbab}} \underset{\text{bab}}{\circ} \xrightarrow{\text{babb}}$$

The last two quadruplets in our seven-letter string, *babb* and *abbb*, overlap in *abb*. Here we display all the overlaps in the string *abbabbb*.

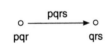

Now let's take a completely fresh view of the quadruplets and their overlaps. Instead of thinking in terms of strings, we now are involved with what we can pretend are one-way roads and towns. The towns are the 8 triplets, and the roads are the 16 quadruplets. The figure is only a small part of a larger system of one-way roads and towns, with the towns (the triplets) recording the overlaps of the roads (the quadruplets).

The quadruplet *pqrs* becomes a road that goes from town *pqr* to town *qrs*. Each triplet occurs as the final triplet of two quadruplets and the initial triplet of two quadruplets.

Letting the letters *p, q, r, s, t, u,* and *v* stand for the letters *a* or *b*, we see what the system of roads and towns looks like near the town *pqr:* Each town is at a junction of four roads, two roads that enter it and two that leave it. For example, the figure at the lower left shows what the highway system looks like near the town *abb*.

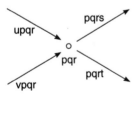

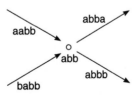

The roads near the town *aaa* look a bit different. One of these roads, *aaaa*, is a loop: This road both enters and leaves the town. The figure below shows the roads near this town. Similarly, the roads near *bbb* also include a loop. The only towns with loops are *aaa* and *bbb*.

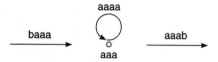

But even at towns *aaa* and *bbb*, two roads enter and two leave.

The entire highway system of 16 roads and 8 towns is shown here:

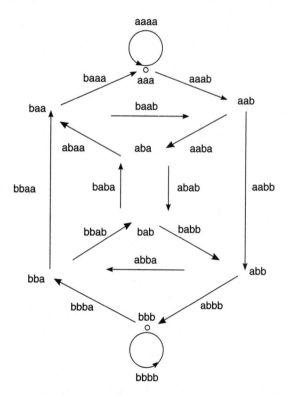

This map of mythical towns and roads displays a great deal of information. It shows all the overlaps of the 16 quadruplets.

Glancing at a particular quadruplet, a road, we can easily read off the 2 quadruplets that could possibly follow it in a full string. For instance, what could follow the quadruplet *babb*? It ends in the town *abb*. Since the roads *abba* and *abbb* leave that town, either one could be the next quadruplet after *babb*. That corresponds to either *babba* or *babbb* appearing in a string.

This fresh way of looking at strings and their overlaps will turn out to be the key to showing that full strings exist not just for couplets and triplets but for quadruplets and so on.

Imagine that you travel about in this highway system, but you never go the wrong way on any one-way road. For instance, you could take a short journey, say, from *bba* to *baa*, passing through the towns *bab* and *aba* along the way. You would travel on the roads

bbab, baba, and abaa

in that order. What does this mean in terms of strings?

The first and second quadruplets (roads) overlap in the triplet (town) *bab*. The second and third quadruplets overlap in the triplet *aba*. We display the three quadruplets and their overlaps in a staircase, as we had done with triplets:

bbab
baba
abaa

Because of their overlaps, all three quadruplets could be recorded in the short string *bbabaa*.

The task of devising a full string that lists the 16 quadruplets can now be rephrased in terms of the highway system shown above. Keep in mind that we want to run through the 16 quadruplets in such a way that adjacent quadruplets overlap in a triplet.

Consider a thrifty highway inspector who wants to check each road for potholes. To save on gasoline, she hopes to plan a trip that takes her over each one-way road in the legal direction once only. Is there such a route? That is how the question "Is there a string that lists all 16 quadruplets once each?" now reads. We have left strings and wheels far behind and are now involved with finding routes through a system of highways. Such systems are studied in the branch of mathematics known as *graph theory*, in which towns are called *vertices* and roads are called *edges*.

Finding a Route

Before we try to find a route for our highway inspector, we should check that the system comes "in one piece." She should be able to get from any town to any other by a legal path. A few minutes' inspection of the map will show that she can. However, we will show that indeed she can do this without looking at a picture. Moreover, our reasoning will apply to the much more complicated highway systems that arise for 5-tuplets, 6-tuplets, and so on.

Say that the inspector is on the road *pqrs* and wants to reach the road *tuvw*. (As usual, the letters stand for the letters *a* and *b*.) The route described in the following staircase will do the trick. It may not be the shortest route that meets her needs, and it may have duplications, but that doesn't matter. It does show that she can get from any road to any other.

> *pqrs*
> *qrst*
> *rstu*
> *stuv*
> *tuvw*

This route is shown here:

pqrs   o   qrst   o   rstu   o   stuv   o   tuvw

qrs   rst   stu   tuv

Let us illustrate this with a specific example in the highway system shown on page 85. Say that the inspector wants to go from the road *bbba* to the road *abba*. Our method would give her the route

> *bbba*
> *bbaa*
> *baab*
> *aabb*
> *abba*

You may trace it out on the map and also find the inspector a shorter route from the road *bbba* to the road *abba*.

We are now assured that the inspector can get from any road to any other (while passing over at most three more roads).

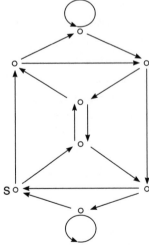

The Inspector Has a Route over All Roads

To find the inspector a route taking her over each road once, let us strip our highway system to its essentials, omitting the names of the roads. The figure at the left displays this stripped-down version, which is all we need for discussing routes. To find a route over all 16 roads, the inspector starts at any town of her choice, say the one labeled S in the figure. She looks at her map and imagines starting to drive at random. Every time she enters a town, she chooses an exit. Since there are just as many exits (2) as entries (2) at each town, she won't get stuck anywhere except when she happens to bump into the town S a second time. (The first time she returns to S, she will find an escape, but not the second time.)

In any case she wanders about until she gets stuck, which will always happen at S. If she is lucky, she may have traveled over all 16 roads. But her trip may have covered fewer roads; for instance, she may have journeyed over just 8 sections, as shown to the right. That leaves her with still 8 roads to inspect. Though she hasn't covered every road, at least she hasn't covered any road twice.

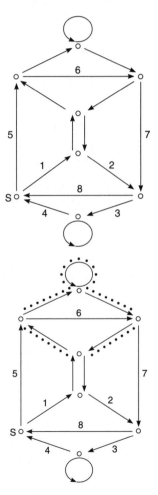

The loop at the bottom she could include as a short side trip between the ones marked 3 and 4. That is, she could alter her trip by covering 1, 2, and 3, then whirl around the loop, then go on to 4, 5, 6, 7, and 8. That still leaves her with 7 roads to cover, but they can be included in side trips also.

Between 5 and 6 she could add a side trip covering 5 more roads, as shown with the dotted lines.

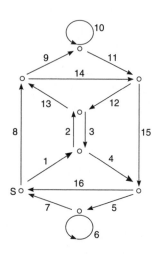

That leaves only the 2 roads in the middle to cover. A short side trip between steps 1 and 2 sweeps them out and the inspector finally has her perfect, economical trip, shown by the numbers in the figure to the left.

Such a complete inspection trip must end at the town where it begins, *S*. All the other towns have as many exits as entrances, but beginning right after she has started her trip, town *S* has only one exit available, though it still has two entrances.

Going back now to our highway system, we see that the inspector has inspected all 16 stretches of roadway in the order shown in this staircase:

1 *bbab*
2 *baba*
3 *abab*
4 *babb*
5 *abbb*
6 *bbbb*
7 *bbba*
8 *bbaa*
9 *baaa*
10 *aaaa*
11 *aaab*
12 *aaba*
13 *abaa*
14 *baab*
15 *aabb*
16 *abba*

You should not be surprised to see that the final quadruplet *abba* overlaps the first quadruplet *bbab* in a triplet—namely, the triplet *bba*, the town *S*. That just reflects the fact that the trip must end where it starts.

Just for the record, this figure shows the wheel for quadruplets that corresponds to the inspector's trip. This wheel sweeps out all 16 quadruplets no matter where you start.

a a b b
b a
a b
a a
a b
a b b b

The Next Question

Now that we have shown that there are full strings for triplets and quadruplets, we should see if there is a full string for listing 5-tuplets. Or, what amounts to the same, decide if there is a wheel that lists all 5-tuplets. It turns out that the work we did with the inspector and the highway system for quadruplets will help us settle the 5-tuplet case with little effort.

Just as there are twice as many 4-tuplets as 3-tuplets, there are twice as many 5-tuplets as 4-tuplets: 2 times 16, or 32 5-tuplets. Reasoning as before would show that any string with more than 32 + 4, or 36, letters in it will have a duplicated 5-tuplet. To show that there is a string of length 36 without duplications, let's do almost exactly what we did with our mythical highway map for 4-tuplets. Not exactly, since the highway system would now have 32 roads and 16 towns. Rather than draw such a complicated arrangement, we will simply describe it.

Each road now stands for a 5-tuplet of *a*'s and *b*'s, *pqrst*. Each such road goes from a town *pqrs* to a town *qrst*. Once again, two roads enter each town and two roads leave it. Moreover, using the same reasoning as before, we could show that the inspector could travel from any road to any other.

Without even drawing the system, we conclude that the inspector has a route that passes over each road exactly once and ends at the town where the trip begins. That tells us, first, that there is a full string for 5-tuplets. Second, it assures us that its final four symbols are the same as its first four symbols. From this information we conclude that there is a wheel with 32 a's and b's in which every 5-tuplet appears exactly once.

We are now in a position to answer the general question, "How long a string exists where no n-tuplet is repeated?"

Recall that for the four couplets we made a string of length 5. For the eight triplets our string had length 10.

Just as there are $2 \times 2 \times 2 = 2^3 = 8$ triplets, $2^4 = 16$ quadruplets, and $2^5 = 32$ 5-tuplets, there are 2^n n-tuplets, for each time we increase n by 1, the number of words of that length doubles. A wheel for all the n-tuplets has 2^n symbols. A string for all the n-tuplets would have to be $n - 1$ symbols longer in order to display all of the final n-tuplet. For instance, when n is 3, then $n - 1$ is 2, and the string has length $8 + 2 = 10$, as we saw.

Filling a Gap

We skipped over an important point in order not to interrupt the flow of our analysis. Now is a good time to take it up. It concerns the side trips. How do we know that the inspector can always add a side trip to a trip if she hasn't yet covered all the roads? Could it happen that in a partial journey from some starting point S back to the same point she has used up all the roads adjacent to every town she has passed through, even though she hasn't covered all the roads? If that can happen, then she will not be able to add a side trip.

Here is how we can show that this will never happen: Imagine that the inspector's partial journey covers all the roads that touch each of the towns she has passed through. Then those towns and the roads joining them form a system from which the inspector cannot escape. That means that no one could drive from any of those roads to reach an uninspected road. But we showed that it is possible to drive from any road to any other road. We conclude that as long as there are uninspected roads, there will always be side trips the inspector can add to her trip.

Yet More General

What was it about the highway system that convinced us that the inspector could always find some route that would allow her to drive over each road only once? First, we used the fact that as many roads enter a town as leave it. Second, the system consisted of one piece: She could reach any road from any other road. That's all, just an assumption about the small piece of a highway system near each town, and an assumption about the whole system forming one connected piece. We have really obtained quite a general insight into a property of certain highway systems. Now let's apply this insight to extend our discoveries about strings that have no duplications.

For instance, what if we formed strings with the letters a, b, and c? To be specific, say that we wanted to show that there is a wheel for 10-tuplets made with the letters a, b, and c. (There are $3 \times 3 \times 3 \times 3 \times 3 \times 3 \times 3 \times 3 \times 3 \times 3 = 3^{10}$ such words, just as there were 2^{10} 10-tuplets made with the letters a and b.) In this case we imagine a highway system in which each road is one of those 3^{10} words. That's 59,049 roads. Luckily we don't need to draw any of them. The

3^9 9-tuplets of *a*, *b*, and *c* will be the towns. That's 19,683 towns. Each road leads from its opening 9-tuplet to its ending 9-tuplet.

How many roads leave each town? For instance, how many roads leave the town *abaccbacc*? That 9-tuplet is the first 9 symbols of 3 (not 2) 10-tuplets since we can add any of the 3 letters *a*, *b*, or *c* to it to make a 10-tuplet whose first 9 symbols are the given 9-tuplet. So 3 roads leave each town. Similarly, 3 roads enter each town.

Since three roads leave each town and three roads enter it, the same number leave as enter. Moreover, just as we showed that the system for quadruplets of *a* and *b* comes in one piece, we could show that the same holds for our much larger system. In fact, an inspector can go from any road to any other road by a route that passes over at most nine intermediate roads. That's surprising, given that there are so many towns and roads.

We can immediately conclude that the inspector of this vast system does have a route that sweeps out each road exactly once. That tells us that there is a wheel with 59,049 symbols that lists all 59,049 10-tuplets of *a*, *b*, and *c* exactly once each.

We reached this conclusion almost without any effort. We didn't even have to draw anything or make any lists. The deeper we went into the original problem for triplets and quadruplets, the more general our thinking became. Clutter gave way to clarity. The concrete was replaced by the abstract, and the abstractions turned out to be extremely useful.

Finally, what if we wanted to design a route for a salesperson rather than a road inspector? A salesperson would like a route that passes through each town exactly once. This turns out to be a much tougher problem. No one has found a simple way to decide whether such a route exists in a given highway system. Checking all possible routes in a system with many towns is not a simple task, even for a high-speed computer.

Applications

Wheels or strings of the type discussed in this chapter have had many and varied applications. One of their more recent uses is long-range radar, such as the type that measures the distance to the moon. Strings for 50-tuplets of a and b are fed in by a computer. Such strings go on for billions of symbols before repeating any 50-tuplet. (A pulse of energy is used for a, and its absence for b.) Because the 50-tuplets are not repeated, it is possible to identify exactly when a transmitted signal, after bouncing off an object, returns.

How are such long strings produced? Not by the method described in this chapter, for it would be much too slow. Instead, algebra is used to provide a formula that gives the symbols rapidly. The formula can easily be programmed for a computer. Some fascinating pure mathematics has been developed to meet the challenge of finding the right formula.

Inside a computer there is a program for spotting and correcting errors in manufacture, known as a *BIST*, or a *built-in self-test*. This program and similar ones to help achieve reliability are also built into CD-ROMs and telephones. Electrical engineers call them *shift register sequences*, and they are constructed with the aid of the strings described in this chapter.

Every time you make a call on your cell phone, you are identified by a personal signature, which is a shift register sequence. This arrangement is known by its acronym *CDMA*, which stands for *code division multiple access*.

Because shift register sequences appear fairly random, they are also used to conceal secret messages. First, a message is translated into a sequence of pulses and no pulses, and then a shift register sequence is mixed into it. The receiver, knowing which sequence had been added, removes it and recovers the original message.

Certainly the first application of these strings occurred in the poetry of India some time between 200 and 2000 years ago. The 8 rhythms of 3 syllables all occur in this memory word of 10 syllables:

ya má tá rá ja bhá na sa la gám

This is just a disguised version of the string *abbbabaaab*. A mark over a syllable stands for a heavy beat. The first rhythm, ya ma′ ta′, is light, heavy, heavy; the next is ma′ ta′ ra′, heavy, heavy, heavy, and so on to the last, which is light, light, heavy.

Throughout this chapter we've examined arrangements of symbols on a line without certain repetitions. A variation on this theme, arrangements of symbols in a rectangle, has played a role in the remote control of robots.

Consider a robot free to move within a 3-by-9 rectangle made up of 27 cells. Each cell contains one of two shapes that the robot is programmed to recognize. Call the shapes *a* and *b*. The robot moves right or left, up or down, but does not rotate. On its base is a 2-by-2 window that can sense four exposed cells at a time and transmit this information to the operator. In order that the operator can tell where the robot is located, all the 2-by-2 blocks in the 3-by-9 rectangle

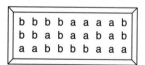

must be different. They are in our rectangle, as you may check.

Of course, if we allow the robot to rotate, then this design is no longer adequate. For instance, the operator couldn't tell if the robot were in the top right corner or the bottom left corner. We will leave such concerns for specialists to face.

As we look back over the chapter, we may ask ourselves what mathematics we used. Not algebra, not geometry, no particular fact that we may have picked up in school. Rather, we depended on just three important tools in the mathematical style of thinking: common sense, careful attention to details, and a willingness to be flexible. No wonder that the study of mathematics provides an ideal preparation for so many careers.

CHAPTER SIX

Counting Ballots

Imagine an election in which there are two candidates, Ann and Barbara. The ballots are in the ballot box and about to be counted one by one. Barbara has more votes than Ann. What is the chance that Barbara stays ahead of Ann throughout the count? If we let the letter *a* stand for a vote for Ann and the letter *b* stand for a vote for Barbara, this question will quickly lead us to investigate strings of *a*'s and *b*'s.

Let's start with a simple case in which there are three *b*'s and two *a*'s. Here is a list of all the possible orders in which the five ballots can be pulled out of the box:

aabbb
ababb
abbab
abbba
baabb
babab
babba
bbaab
bbaba
bbbaa

You could make your own list to check that this list is complete. There are 10 possible orders.

If an *a* is drawn at the very start, Ann is ahead. Then there is no chance that Barbara leads all the time.

If a *b* appears first and then immediately an *a*, Ann and Barbara are tied. Again there is no hope for Barbara to lead throughout, no matter in what order the remaining ballots are drawn, since no one leads in a tie. If Barbara is going to be ahead all the time, then she must get the first two votes pulled from the ballot box.

These two observations eliminate the first seven strings in the list. In the eighth, *bbaab*, Barbara loses the lead at the fourth ballot counted. Only in the final two strings, *bbaba* and *bbbaa*, does Barbara always maintain a lead.

Since in 2 of the 10 possible orders of counting the ballots, Barbara always stays ahead, we say that the likelihood of her retaining the lead throughout the count is 2 out of 10, or 2/10, or 20 percent.

With many more votes—hundreds or thousands—a list of the possible orders would be so long we would not want to write it down. If we felt ambitious, we might assign the task to a computer, but even the computer may be reluctant to face so many possible strings and check in which ones the *b*'s always lead the *a*'s.

Of course we could save time by not bothering to write any string that starts with an *a*. In our list we can easily see that 4 of the 10 strings begin with *a*. That is 4/10, or 40 percent of the strings. Sixty percent begin with *b*. It is among these latter strings that we find the strings in which Barbara leads all the way.

In the case in which Ann has 10,000 and Barbara 15,000 votes, there are 25,000 ballots in the box. We may not list all the possible orders, but we can at least find the fraction of the orders that begin with *a*. In those orders Barbara cannot possibly lead all the way.

If you reach into the ballot box without peeking, you have 10,000 opportunities to grab an *a* and 15,000 of grabbing a *b*. That means the likelihood of plucking an *a* at the start is 10,000 out of 25,000, or 10,000/25,000, which reduces to 10/25, and finally to 2/5, or 40 percent. To put it simply, "Two fifths of all the possible orders of counting the ballots start with a vote for Ann."

We don't know how many possible orders there are, but we do know that 2/5 of them start with an *a* and 3/5 of them start with a *b*. We could pause to figure out the fraction that start with *bb*. And then we could calculate the odds of the first three votes being *bbb* or *bba*. Matters would quickly get very complicated as we calculate the odds that Barbara is ahead throughout the first four votes, then for the first five. We would give up because the work would exhaust us.

Happily, there is a neater solution, another "one-line solution," though the line may occasionally be rather long. Sometimes a one-line solution may require a paragraph; nevertheless, a one-line solution should be elegant and, in retrospect, simple. This whole chapter is devoted to presenting the one-line solution of our voting problem. Once you read over the solution, you will, I hope, agree that it could have been expressed in a few words, rather tersely perhaps, but clearly summarized. As we will see, it depends on one delightful geometric insight. I call it "delightful" for it solves our problem in a blinding flash.

Collecting Data

At this point, when we still have no idea what the answer may be, we do what mathematicians usually do in such situations. We experiment, starting with the simplest cases. We will do that, record the data, and hope to find a pattern. With luck, the data may even suggest a way to analyze the problem.

In order to discuss the cases, we introduce some notation. The number of votes for Ann we will call $N(a)$, which we read as "N of a." $N(b)$ is the number of votes for Barbara. In our opening example $N(a)$ is 2 and $N(b)$ is 3. We also mentioned the case where $N(a)$ is 10,000 and $N(b)$ is 15,000.

The simplest case is when Ann gets no votes at all—that is, when $N(a)$ is 0. In that case Barbara is always ahead in the count. In numerical terms, there is a 100 percent chance that Barbara stays ahead throughout the count.

The next case occurs when Ann gets exactly one vote, that is, when $N(a)$ is 1 and $N(b)$ is therefore at least 2, since we always are assuming that Barbara wins the election. Of all such cases, the simplest is the one in which Barbara gets just two votes. Here is the list of all ways the total of the three votes can be counted:

abb
bab
bba

Only in the last of the three cases does Barbara stay ahead throughout the count. In short, her likelihood of staying ahead is 1 out of 3, or 1/3.

The next case has $N(a)$ equal to 1 and $N(b)$ equal to 3. There are now four possible orders of the counting:

abbb
babb
bbab
bbba

Only in the last two cases does Barbara lead all the way. The likelihood of this is then 2 out of 4, which is 2/4.

The next case has $N(a)$ equal to 1 and $N(b)$ equal to 4, for which there are five possible outcomes:

> *abbbb*
> *babbb*
> *bbabb*
> *bbbab*
> *bbbba*

In this instance the likelihood of Barbara staying ahead is 3/5.

You may want to run through a few more cases where $N(a)$ is 1, just to be sure you have the feel of the problem and data.

So far, this is the information we have gathered, including the case in which $N(a)$ is 2 and $N(b)$ is 3:

$N(a)$	$N(b)$	Likelihood
1	2	1/3
1	3	2/4 = 1/2
1	4	3/5
2	3	1/5

There's not much evidence here on which to base a conjecture. You may want to check the case in which Ann has two votes and Barbara has four votes. But we will check the next case in which $N(a)$ is 2 and $N(b)$ is 5. There turn out to be 21 possible strings or orders, listed here:

> *aabbbbb*   *bbaabbb*
> *ababbbb*   *bbababb*
> *abbabbb*   *bbabbab*
> *abbbabb*   *bbabbba*

abbbbab	bbbaabb
abbbbba	bbbabab
baabbbb	bbbabba
bababbb	bbbbaab
babbabb	bbbbaba
babbbab	bbbbbaa
babbbba	

To convince yourself that the list is complete, look carefully at the two *a*'s. They methodically run through all their possible locations, sweeping out the left column first. In 9 of the 21 cases, Barbara holds her lead throughout the count. Our summary list now reads:

N(a)	N(b)	Likelihood
1	2	1/3
1	3	2/4 = 1/2
1	4	3/5
2	3	2/10 = 1/5
2	5	9/21 = 3/7

Before reading on, spend awhile inspecting these data. What seems to be a formula for the fraction? You will probably spot the simple pattern. (To see this pattern, you will sometimes have to use the unreduced fraction, sometimes the reduced one.)

It seems, judging by the information gathered so far, that the numerator of the likelihood fraction is the difference between $N(b)$ and $N(a)$—that is, $N(b) - N(a)$—and that the denominator is the sum of those two numbers, $N(b) + N(a)$. In short, the likelihood of Barbara's staying ahead is simply

$$\frac{N(b) - N(a)}{N(b) + N(a)}$$

That is an educated guess, not an established fact.

Further Checks

You may want to check our conjecture by running some more cases, for instance, when Ann has two votes and Barbara has six votes. In any event, it would be wise to stop and do another case just to get a deeper feeling for the problem. Another check is to see what the formula tells us when Ann gets almost as many votes as Barbara. In this case the odds of Barbara staying ahead throughout the count should be fairly small. Does the formula support this assertion? When Ann gets almost as many votes as Barbara, $N(b) - N(a)$ is small. Therefore, the numerator in our formula is small and the whole fraction is therefore small. That agreement with our intuition gives us more confidence that the formula is correct.

We can also check our conjecture experimentally. We can count the ballots one by one many times and see in what fraction of the cases Barbara stays ahead. For instance, take the case in which Barbara has 7 votes and Ann has only 3. We may use 7 red checkers to stand for Barbara's votes and 3 black checkers to represent Ann's 3 votes. If you don't have checkers, poker chips will do. In any event, put the 10 "votes" in a jar, shake madly, and pull the votes out one by one without peeking. Record whether or not the b's stay ahead all the time. Then do the experiment again and again until you have accumulated lots of data. In each case, if the a's ever catch up or lead, you can stop the count.

I did this experiment 50 times with poker chips. Barbara stayed ahead of Ann in 21 of the counts. How does this compare with our conjecture? In this case, $N(a)$ is 3 and $N(b)$ is 7. The conjecture predicts that the likelihood that Barbara will hold the lead throughout the count is

$$\frac{N(b) - N(a)}{N(b) + N(a)} = \frac{7 - 3}{7 + 3} = \frac{4}{10}$$

In short, Barbara should stay ahead in 40 percent of the times that the count is made. Since I conducted the experiment 50 times, and 40 percent of 50 is 20, we would expect about 20 of the trials to have Barbara always ahead. The experimental result of 21 out of 50 trials is so close to the theoretical 20 that it gives further evidence that the conjecture may be right.

Another check is to see what the conjecture says when Ann gets no votes at all, that is, when $N(a)$ is 0. In that case the formula becomes

$$\frac{N(b) - 0}{N(b) + 0}$$

which reduces to 1, or 100 percent. That agrees with our observation earlier that when Ann gets no votes, Barbara always has the lead throughout the count.

One more remark before we try to see whether the formula is correct. If Barbara leads Ann during the entire count, the very first ballot picked must be for Barbara. What is the likelihood that this happens?

In the ballot box are $N(b) + N(a)$ ballots. Of these, $N(b)$ are for Barbara. So the odds of picking out a ballot for Barbara are

$$\frac{N(b)}{N(b) + N(a)}$$

Now, our conjecture that the likelihood that Barbara stays ahead the whole count is

$$\frac{N(b) - N(a)}{N(b) + N(a)}$$

This fraction is less than the fraction $N(b)/[(N(b) + N(a)]$. It should be since "staying ahead the whole count" is less likely than "getting the first ballot picked." That is further evidence in favor of our conjecture.

Searching for an Attack

The conjectured formula is so short and simple that there ought to be a short and simple explanation for it—especially if it is correct.

One way to try to explain it is long and complicated: First, calculate the total number of orders in which the ballots could be counted. In other words, work out a formula for the length of the list of possible outcomes in terms of the numbers $N(a)$ and $N(b)$. In the examples we treated, this was easy because the lists were so short. Then we would try to calculate the number of these strings in which Barbara leads all the way. Dividing this second number by the length of the list would give us the likelihood that Barbara stays ahead during the whole count. Such a tedious approach is not attractive, especially in view of the simplicity of the formula we are hoping to justify.

Another approach is to translate the problem into geometrical terms. That's what we did in the preceding chapter, where we began with a problem about not duplicating certain words and ended up driving around highway systems.

A Bug Enters the Picture

This time let us imagine a bug walking about on an endless line. The line is marked like a ruler:

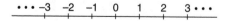

The bug starts at 0. As the ballots are counted, he moves 1 unit to the left or right depending on whether the ballot is for Ann or for Barbara. When an *a* is pulled out, he moves to the left 1 unit. When a *b* is pulled out, he moves to the right 1 unit.

Say that the ballots are drawn in the order *bbaba*. Then the bug starts by moving to the right twice, then once to the left, then once to the right, and finally once to the left. His successive positions are shown in the following figures as heavy dots.

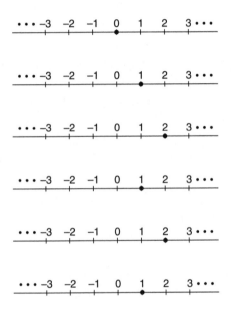

In this particular case the bug stays to the right of his starting position, 0. That is the pictorial equivalent of saying, "Barbara stays ahead of Ann throughout the count."

We can now state our general ballot-counting problem geometrically. A bug wanders on a line, starting at 0. In some random order he takes $N(a)$ steps to the left and $N(b)$ steps to the right, where $N(b)$ is larger than $N(a)$. What is the likelihood that the bug stays to the right of 0 throughout his entire journey?

Unfortunately, diagrams like the ones we have just seen give us no deeper insight. They consist of many separate figures, one for each ballot cast. The whole count is not shown in one picture.

There is another geometric approach, an approach that displays the entire count in one figure. This approach also makes use of a bug, but this bug crawls about in a plane, not just on a line.

We draw a pair of lines at a right angle to each other and mark them like a ruler. The horizontal line has the numbers 0, 1, 2, 3, . . . , while the vertical one also has those numbers as well as the negative numbers −1, −2, −3, . . . , as is shown below:

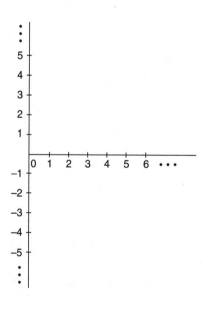

The bug starts at the point where the two lines meet. When a vote for Barbara is drawn, he moves 1 unit to the right and 1 unit *up*. When a vote for Ann is drawn, he moves 1 unit to the right and 1 unit *down*. These two moves are shown here. In both cases the bug moves to the right. The bug's path thus consists of line segments tilted upward or downward.

For example, if the votes are counted in the order *bbaba*, the bug moves upward twice, then downward once, then upward, and finally downward, as is shown in the diagram below:

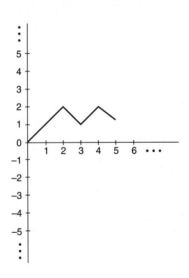

Now the entire vote count is visible at a glance. That the bug's path remains above the horizontal line indicates to us that Barbara has maintained a lead throughout the entire count.

In any count in which Barbara is always ahead, the first ballot counted must be b. That means the first move of the bug must be upward, like this:

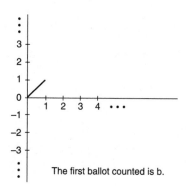

The first ballot counted is b.

As we saw, the chance of the first ballot counted being b is given by the fraction

$$\frac{N(b)}{N(b) + N(a)}$$

The likelihood that the first ballot is an a is

$$\frac{N(a)}{N(b) + N(a)}$$

In that case the first step of the bug is downward, like this:

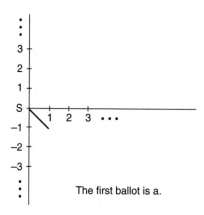

The first ballot is a.

The Geometrical View

When Barbara gets $N(b)$ votes and Ann gets $N(a)$ votes, what does the bug's path look like, even if Barbara doesn't stay ahead in the count?

First, the bug starts at the point S where the horizontal and vertical lines in the figure meet. Then the bug crawls about, but with each step it goes 1 unit to the right. So it ends up $N(b) + N(a)$ steps to the right. During his journey the bug goes up $N(b)$ times and down $N(a)$ times. Since $N(b)$ is larger than $N(a)$, the bug ends up above the horizontal line—in fact, $N(b) - N(a)$ units above it. So we can draw the point where his journey ends. It is the point labeled F, for "finish," in this figure:

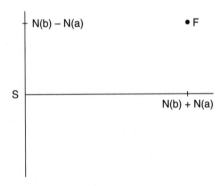

The bug has many possible paths from S to F, depending on the order in which the ballots appear during the count. We want to compute the fraction of those paths that stay above the horizontal line. Based on skimpy data, we suspect that this fraction is

$$\frac{N(b) - N(a)}{N(b) + N(a)}$$

However, at this point we are not sure our suspicion is correct.

A Clue

The fraction we just displayed can be written as the difference of two fractions:

$$\frac{N(b)}{N(b) + N(a)} - \frac{N(a)}{N(b) + N(a)}$$

As we saw earlier, the first of these two fractions represents the likelihood that the first vote counted is b. Geometrically speaking, this means that the bug starts his journey with an upward step. The second fraction represents the likelihood that the first vote is an a, and the bug moves downward at the start.

Barbara stays ahead of Ann throughout the count if the bug starts upward and never touches or crosses the horizontal line before he reaches F, the end of his path. To find the likelihood of wandering in such a path, we start with the likelihood that the path starts with an upward step and then subtract from it the likelihood that it then goes on to meet or cross the horizontal line. We are interested in the difference

$$\frac{N(b)}{N(b) + N(a)} - \text{the likelihood that the path starts upward and meets the horizontal line}$$

Look closely at the last two displays. They both begin with the same fraction. If we can show that the two expressions that are subtracted are equal, our conjecture is true. Recall that $N(a)/[(N(b) + N(a)]$ is the likelihood that a path begins with a downward step. So: "*Why is the likelihood that a path starts with an upward step and then meets or crosses the horizontal line equal to the likelihood that a path starts with a downward step?*"

The Key to Everything

The question is: "*Why is the number of paths that start with an upward step and meet or cross the horizontal line equal to the number of paths that start with a downward step?*"

At first glance we wouldn't expect any relation between the two types of paths. Yet, if our conjecture is correct, then the two numbers must be equal. That is, the number of paths that start with *b* and meet or cross the horizontal line equals the number of paths that begin with *a*. I suggest that you experiment with the case that $N(a)$ is 2 and $N(b)$ is 3 and see if the two numbers are equal before reading further. The left column in the figure below shows the four paths that start with an

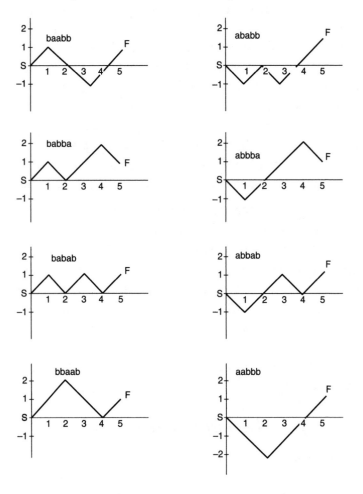

upward step and meet the horizontal line; the right column shows the paths that start with a downward move.

Is there some easy geometric way to obtain the four paths in the right column from the four paths in the left column? We hope that there is some method and that the method works no matter how many votes Barbara and Ann receive, as long as Barbara gets more than Ann does.

One tempting possibility is to reflect each path in the left column around the horizontal line. In other words, flip it around that line. That certainly turns a path that starts with an upward step into a path that starts with a downward step. Unfortunately it also turns the finish point of the paths F into a point other than F, namely, a point below the horizontal line. A good try, but no cigar. However, a slight change in the reflection in the horizontal line will work.

The Modified Reflection

We didn't use the fact that the paths we are interested in not only start with an upward step but that they touch or cross the horizontal line at least once at a point other than the starting point. Call the first such point C. The figure below

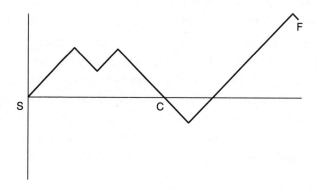

displays this point for a typical path that has those properties for any election in which Barbara has more votes than
Ann.

Instead of reflecting the whole path around the horizontal line, we can reflect *just the part from S to C*, leaving
the part from C to F unchanged. The reflection, together
with the unchanged part from C to F, is a path that starts
with a downward step and ends at F. It is shown dashed in
the figure below.

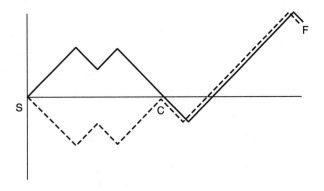

This restricted reflection does the trick. Going back to our
two columns of paths, the case in which Barbara has 3 votes
and Ann has 2, we can see that the paths in the right column
are obtained from the similarly placed paths in the left column by this reflection technique. Our reflection method
applies in general, not just to the special case shown.

Every path from S to F that starts with a downward step
can be obtained this way—regardless of the total number of
ballots. To see why, note that such a path must touch the
horizontal line at least once. (If Ann starts ahead in the
count and ends up behind, at some point she will be tied

with Barbara.) Call the first point where the path touches that line C. Reflect the part of the path from S to C around the horizontal line to obtain a unique path that starts with an upward step and first touches the horizontal line at C. Reflection of the part of that path from S to C gives us the path we started with.

We now know that the number of paths from S to F that start with b and touch the horizontal line is the same as the number of paths from S to F that start with a.

It follows that the likelihood that Barbara stays ahead of Ann throughout the whole count is

$$\frac{N(b)}{N(b) + N(a)} - \frac{N(a)}{N(b) + N(a)}$$

Putting both of these fractions over their common denominator gives us our conjectured formula:

$$\frac{N(b) - N(a)}{N(b) + N(a)}$$

Our educated guess is right: The likelihood that Barbara keeps her lead throughout the count equals the difference between the number of votes for her and the number of Ann's votes divided by the total number of ballots.

Could we have discovered this fact without resorting to pictures? Perhaps. This is how the reasoning would go if we work directly with the strings of a's and b's.

Consider a string that begins with b and at some point has a tie. In the section that ends with the first tie, replace every a with b and every b with a. These switches don't change the number of a's or the number of b's. However, they produce a string that begins with *a*. This is exactly what the geometric reflection accomplishes.

Some Reflections on That Reflection

Is it fair to call our reasoning a "one-line argument"? Since it took several pages, our first impulse is to say no. However, if we had omitted the examples and the guiding commentary, we would have had a pretty short, but dense, explanation. This chapter would shrink to a page or two. The key to the whole chapter is the reflection principle that is illustrated in the last figure. Perhaps we should have just displayed that figure and described how it pairs off paths of one type with paths of another type. Then we could have mentioned the ballot-counting problem as an application. The reasoning boils down to one picture, if not to one line.

Another Application of Reflection

The use of the "reflection in a line" may seem like a trick pulled from the clouds, but this is not the case. Reflection is a standard tool in mathematics, almost as handy as a hammer is to a carpenter. To convince you, I'll give another example of its use.

A hiker at point *A* in the figure below wants to get a pail of water from the river and carry it to point *B*. What is the shortest path that she can take?

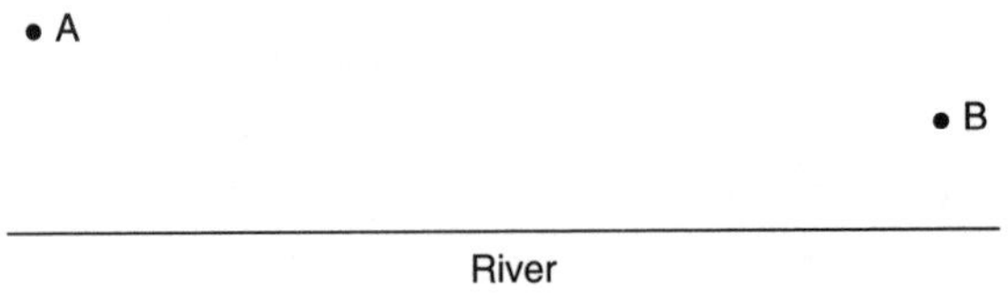

• A

                                                    • B

River

The hiker has many possible paths, three of which are shown here:

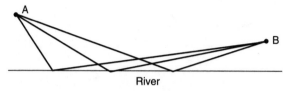

Three paths from A to river to B

To find the shortest path that meets her desire, first reflect *B* across the river and call the new point *C*, as below:

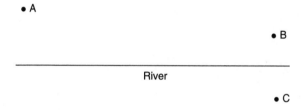

The shortest path from *A* to *C* is simply the straight line shown in the next figure. It crosses the river at a point that we label *D*.

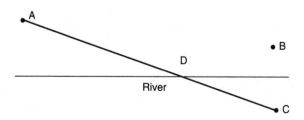

Now, to find the shortest path from *A* to *B*, reflect the part from *D* to *C* around the river, as shown below. This reflection is a straight path from *D* to *B*.

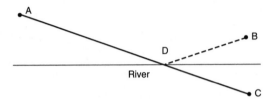

The path from *A* to *D* to *B* is the shortest path from *A* to *B* that includes a visit to the river. To see why, look at any other path, the typical one being *A* to *E* to *B*, where *E* is any point on the river, as shown here:

A typical path

Then reflect the part from *E* to *B* around the river, obtaining a path of the same length from *A* to *C*:

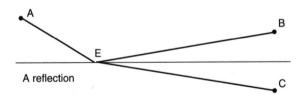

A reflection

If *E* is not the point *D*, the path from *A* to *E* to *B* is equal in length to the path from *A* to *E* to *C*, which is longer than the straight path from *A* to *C*, which crosses the river at *D*.

Therefore, the hiker should go to point D to achieve the shortest walk. This figure summarizes these observations:

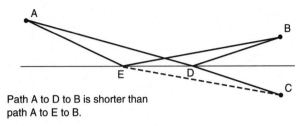

Path A to D to B is shorter than
path A to E to B.

Incidentally, physicists meet the *hiker-and-pail problem* in the study of optics: What path does a light ray take if it starts at A, bounces off a mirror (the river), and reaches B? The answer is, "The route the hiker follows." It's as if the light ray has a personal desire to minimize the time to travel from A to B while bouncing off the mirror. Physicists describe this route by saying that "the angle of reflection equals the angle of incidence," as shown here:

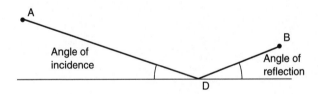

When you look back at the fourth and fifth figures in this last section, which show the path D to C reflected as path D to B, you will see why the two angles are equal.

Once again we have traveled far from our starting place, strings composed of two letters. This is to be expected since there are no barriers separating the hundreds of fields of mathematics from each other or mathematics from the many fascinating phenomena of the world in which we live.

Infinity

The most basic idea in mathematics is that of a "set." A *set* is any collection of objects, which are then called *elements of the set.* The idea is so fundamental that we meet it in kindergarten or first grade. During the 1960s, in the era of the New Math, even the word itself was introduced that early. Maybe teachers don't use the word anymore, but they certainly convey the idea, usually by drawing a loop around a picture of two or three apples or bananas.

In daily life there are many synonyms for the word *set,* depending on the nature of its elements. We speak of a "flock of sheep" (not a "set of sheep") and a "herd of cows." The word *set* is used in tennis; it consists of several games. The elements of the "social set" are the wealthy people who participate in debutante balls and fancy charity benefits.

In school we used *finite sets* to introduce whole numbers, which shows that the idea of a set precedes that of a number. (A set is finite if its elements can be counted—1, 2, 3, . . . ,— and the counting stops.) In this chapter we focus on *infinite sets,* that is, *the sets that are not finite,* which are far more exciting than finite sets, as they offer paradoxes that violate our intuition and quickly lead to contradictions that threaten our faith in common sense. Nothing in our experience with finite sets prepares us for the surprises that infinite sets offer.

We start with a seemingly simple question.

A Simple Question

We will distinguish two types of strings made with the letters *a* and *b,* the finite and the infinite. A finite string stops. The string *ababb* is finite. A finite string can be short, perhaps consisting of just one letter, or as long as we please. There are an infinite number of finite strings. An infinite string goes on forever, such as the string *ababab* . . . , in which *a*'s and *b*'s alternate without end. There is an infinite number of infinite strings as well.

We now define two infinite sets whose elements are strings. One consists of all of the finite strings made with *a*'s and *b*'s. Call this set *F,* for *finite.* The elements of the other set are the infinite strings, the strings that do not end. Call this set *I,* for *infinite.*

As we said, both sets, *F* and *I,* are infinite, and this raises the question: are they equally infinite?

Our first reaction may be to say, "This is a ridiculous question. After all, infinite is infinite. Of course those two sets are equally infinite." That was the attitude until 1873 when George Cantor (1845–1918) took the question seriously and answered it. His answer sent a shock wave through the mathematical community.

Back to Kindergarten

To understand Cantor's discovery, we must first review a concept we learned in kindergarten. I am referring to the notion of "same-size sets," which underlies the definition of the whole numbers, 1, 2, 3, ...

So let's pause for a brief refresher as we, in a sense, become children again.

Here we have a set of trucks and a set of bicycles:

Even before we understood the concept of "two" we could have shown that these sets have "the same size." All we need to do is draw lines or curves pairing each truck with a bicycle. In other words, we marry the elements of one set with the elements of the other set: Each element in one set is uniquely paired with one and only one element in the other set. All the elements in both sets get paired. That is what we mean by "having the same size." The next figure is like dozens we may have seen in kindergarten. It shows that the two sets have the same size:

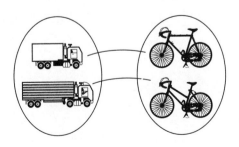

However, there is no way to pair off either of those sets with the set of shoes in this figure:

No matter how you try to arrange the pairing, one shoe will be left stranded.

Throughout this chapter we will use the phrase "same size" in the precise sense we described. It does not mean "same length" or "same area." For instance, the two line segments AB and CD in the figure to the left have the same size but not the same length. To show that they have the same size in our sense, pair off each point P on the segment AB with the point Q on CD that lies on the same ray through the point E as P does.

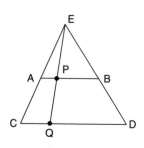

We can apply the definition of two sets being the same size to infinite sets without changing a word. For instance, consider the two sets below. One consists of all the whole numbers and the other of just the even numbers.

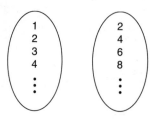

We can pair each whole number with the even number twice as large:

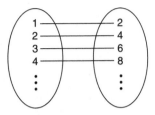

Having set up the pairing, we are permitted to assert that the set of whole numbers has the same size as the set of even numbers. Even though we removed all the odd numbers from the set of whole numbers, what remains has the same size as the original set. This is quite a contrast with the behavior of finite sets. When we take away an element from a finite set, we no longer have a set of the same size. Instead, we obtain a "smaller" set.

What do we mean by *smaller?* We say that one set is smaller than another if it can be paired with part of the second set but it cannot be paired with all of that set.

We will stick to this definition of *same size* as we face the question, "Are the sets *F* and *I* the same size?" That question is really asking, "Is there a way to pair off all the finite strings of *a*'s and *b*'s with all the infinite strings of *a*'s and *b*'s?" If they are not of the same size, then we should decide which is smaller.

Practice

Before trying to answer the question, let us show that the set *F* of finite strings has the same size as the set of whole numbers, 1, 2, 3, ... We are obligated to devise a way to pair off the two sets, element by element, to arrange a mass marriage of the whole numbers with the finite strings.

It turns out that we can arrange such a marriage fairly simply. First, arrange the finite strings in order of their lengths, and then, within a given length, in dictionary order. The list begins:

a, b, aa, ab, ba, bb, aaa, . . .

This suggests how to marry the finite strings to the whole numbers. Just match them, counting from left to right:

a,	*b,*	*aa,*	*ab,*	*ba,*	*bb,*	*aaa,*	. . .
↑	↑	↑	↑	↑	↑	↑	
1	2	3	4	5	6	7	. . .

The pairing continues forever, and we can now say that the two sets have the same size. In short, the set *F* has the same size as the set of whole numbers.

Rephrasing the Problem

If the set *I* of infinite strings were the same size as the set *F* of finite strings, then it, too, could be paired with the set of whole numbers. The following figure suggests why this is so:

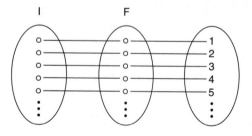

We could just erase the set *F*, so to speak, and carry the pairing straight across from *I* to the set of whole numbers. In other words, we'd pair each infinite string, *s*, with the number that is paired with the finite string paired with *s*, thus:

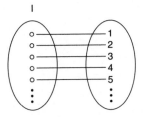

In short, if *I* and *F* have the same size, then so do *I* and the set of whole numbers.

So we face this question: "Can the set of infinite strings be paired with the set of whole numbers?"

Let's begin by showing that the set *I* is at least as large as the set of whole numbers. Here is a pairing of all the whole numbers with some, but not all, of the infinite strings:

$$1 \Leftrightarrow abbbbbbbbb \ldots$$
$$2 \Leftrightarrow aabbbbbbbb \ldots$$
$$3 \Leftrightarrow aaabbbbbbb \ldots$$
$$4 \Leftrightarrow aaaabbbbbb \ldots$$
$$\ldots \quad \ldots$$

The whole number 1 is paired with the string that starts with one *a* and then continues with just *b*'s. The number 2 is matched with the string that starts with two *a*'s and continues with *b*'s; 3 with the string that starts with three *a*'s, the rest being *b*'s, and so on. This shows that *I* is at least as large as the set of whole numbers.

The Answer

We are really asking, "Is there a complete list of the infinite strings of *a*'s and *b*'s such that there is a first string, a second string, a third string, and so on?" If the answer is yes, such a list would look like this, where each dash stands for either an *a* or a *b*.

```
1  — — — — — — — — — ...
2  — — — — — — — — — ...
3  — — — — — — — — — ...
4  — — — — — — — — — ...
5  — — — — — — — — — ...
        ...
```

Every infinite string of *a*'s and *b*'s is supposed to appear somewhere on the list.

George Cantor raised essentially the same question—that is, whether there is such a list of infinite strings—and in a letter of November 29, 1873, asked his friend, Richard Dedekind (1831–1916), whether he could answer it. Dedekind wrote back saying that he could not. On December 2, Cantor replied:

> I was extremely pleased to receive today your answer to my latest letter. I proposed my question to you for the following reason. I had asked it several years ago and had always remained in doubt whether the difficulty which it presents is a subjective one or whether it is inherent in the substance. As you write to me that you too are unable to answer it I may assume the latter. Besides, I would like to add that I have never seriously thought about it because it has no particular practical interest for me and I fully agree with you if you say that for this reason it doesn't deserve too much labor. Only it would be a beautiful result.

A few days later, on December 7, he wrote Dedekind again:

> Recently I had time to follow up a little more fully the conjecture which I mentioned to you; only today I believe

I have finished the matter. Should I have been deceived, I would not find a more lenient judge than you. I thus take the liberty of submitting to your judgment what I have written, in all the incompleteness of a first draft.

On December 7, 1873, Cantor had proved that I, the set of infinite strings in a and b, and the set of whole numbers do not have the same size, and with this discovery he inaugurated the theory of infinite numbers. In his argument he used properties of the number system. But later, in 1890, he devised a much simpler argument, and it this one that I will present.

Imagine any list of infinite strings using the letters a and b, with a string matched with each whole number.

Cantor uses the list itself to construct an infinite string of a's and b's that cannot be on the list. To do this, he begins by circling the letters that appear along the diagonal, as shown below to the right. (Keep in mind that each dash stands for a letter, either a or b.)

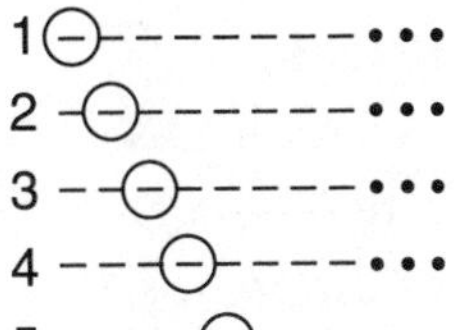

Then he forms an infinite string out of the letters he has circled, starting at the top left corner. This string may or may not appear somewhere in the list.

His next step is the key to what is nowadays called the *diagonal method*. He operates on this string by changing

1 ⊖ – – – – – – • • •
2 –⊖– – – – – – • • •
3 – –⊖– – – – – • • •
4 – – –⊖– – – – • • •
5 – – – –⊖– – – • • •
• • • • • • • • • • •

every a to a b and every b to an a. It is this string, which he derives by the switches, that he claims cannot appear anywhere in the list. Call this string *Cantor's string*. If you think about it for a couple of minutes, you will probably see why Cantor's string can't be any of the strings in the list.

Could Cantor's string, the one he formed by switching the letters along the diagonal, be the first string in the list? No, because it differs from that string at its very first letter, the one that he had circled. (Cantor's string may also differ from that first string at other places as well.)

Could Cantor's string coincide with the second string in the list? No, for it differs from the second string at least in its second letter.

Could Cantor's string happen to be the third string in the list? No, for it differs from the third string at the third letter.

Similarly, Cantor's string cannot be the fourth, the fifth, or the sixth string, and so on. In short, the string he forms by switching letters along the diagonal cannot appear anywhere in the list.

Because the new string Cantor makes differs from any in the list, that list is not complete. Therefore, there is no possible way to pair off the whole numbers with the set I of infinite strings. It follows that the set of whole numbers and the set I do not have the same size, even though both sets are infinite. Therefore, the sets F and I are not the same size, as we set out to prove. In fact, I is larger than F, by an earlier remark.

With this discovery, Cantor founded the theory of infinite sets, eventually showing that there is an infinite number of different sizes of infinite sets. Both this discovery and his diagonal technique have played a major role in the twentieth century in pure mathematics, logic, and computing theory. For example, the logician Kurt Gödel (1906–1978) in 1931, using an argument reminiscent of Cantor's diagonal procedure, showed that there are mathematical truths in any mathematical system containing ordinary arithmetic that cannot be proved within that system. The mathematician Alan Turing (1912–1954) used a variant of the same procedure to show in 1936 that there could never be a universal computing machine capable of duplicating the computations of every possible computer.

One question Cantor raised has inspired some of the deepest research in the foundations of set theory. It can be phrased in terms of the set I of infinite strings, though it usually is expressed in terms of the set of all numbers. This is his question: "Is every infinite set of infinite strings the same size as the entire set of infinite strings or else the same size as the set of whole numbers?" Or, what amounts to the same thing, "Is there a level of infinity between the size of the set of whole numbers and the size of the set I of infinite strings of a's and b's?" It turns out that the answer depends on the assumptions one makes about the basic properties of sets, and research is still being done on his question.

A Paradox

The notion of a set, which appears to be so simple and primitive that we really don't bother to define it, is more complicated than we may expect. For instance, Cantor's discovery that infinite sets come in different sizes challenges our intuition and beliefs even if we follow every step of his diagonal argument and agree that there is no flaw in his reasoning.

We turn now to an episode that took place in 1902, which shows the risks one runs when working with infinite sets. Recall that every set, finite or infinite, is composed of its elements. For instance, the set of even whole numbers has the elements 2, 4, 6, 8, 10, 12, ...

The set of states in the United States has 50 elements. One of these elements is Wyoming. Wyoming is said to "belong to the set of states" or "Wyoming is an element of the set of states."

Another example of a set is "the set of all things that are not shoes." This set is not a shoe. So it is an element of itself. It is certainly unusual for a set to be an element of itself; thus, a set that is an element of itself we will call *unusual*. If it is not an element of itself, we will call it a *usual set*. The set of states is an example of a usual set, since it is not a state. The set of even numbers is a usual set since it is not an even number—it is not even a number.

Bertrand Russell (1872–1970) showed that strange things happen when we consider the set whose elements are all the usual sets. Call this set *S*. Is *S* itself usual, or is it unusual? Surely it has to be one or the other. Let's see which type it is.

Could *S* be a usual set? If so, it is an element in the set of all usual sets. But *S* is that set of all usual sets, so *S* is an element of *S*. That means *S* is an unusual set, for it is an element of itself. Since *S* is now both usual and unusual, the assumption we made at the start must be wrong. That was the assumption that *S* is a usual set.

So it seems that *S* is an unusual set. That means it is an element of itself, that is, it is an element of *S*. But the elements of *S* are the usual sets. So *S* must be usual. Once again we have nonsense, a set being both unusual and usual.

Whether we assume *S* is usual or we assume *S* is unusual, we get a clear-cut contradiction. Yet all we did was define a certain set, *S*. This is known as *Russell's paradox*.

On June 16, 1902, Russell wrote the logician Gottlob Frege (1848–1925), describing his paradox. As he put it, "There is no class (as a totality) of those classes which taken as a totality, do not belong to themselves. From this I conclude that under certain circumstances a definite collection does not form a totality." Here "class" is another word for "set." Russell is saying that though we can define certain vast sets, we may not treat them as casually as we do simpler sets.

Frege responded on June 22, "Your discovery of the contradiction caused me the greatest surprise and, I would say, consternation, since it has shaken the basis on which I intended to build arithmetic." ("Build arithmetic" is short for "provide a foundation for all of mathematics.")

On June 25, Russell wrote his wife, "I have heard from Frege, a most candid letter: he says my conundrum makes not only his Arithmetic, but all possible Arithmetics, totter."

At the same time Frege added an appendix to a book of his which was already at the printer. Its opening paragraph was, "Hardly anything more unfortunate can befall a scientific writer than to have one of the foundations of his edifice shaken after the work is finished." He went on, "This was the position I was placed in by a letter of Mr. Bertrand Russell, just when the printing of this volume was nearing its completion."

Many years later, in 1962, when Russell was asked for permission to publish his letter to Frege, he replied:

> As I think about acts of integrity and grace, I realize that there is nothing in my knowledge to compare with Frege's dedication to truth. His entire life's work was on the verge of completion, . . . his second volume was about to be published, and upon finding that his fundamental assumption was in error, he responded with intellectual pleasure clearly submerging any feelings of personal disappointment. It was almost superhuman and a telling indication of that of which men are capable if their dedication is to creative work and knowledge instead of crude efforts to dominate and be known.

Various mathematicians have patched up the axioms of set theory in ways that avoid the pitfall of Russell's paradox. In one method it is postulated that there are two types of sets: those that are elements of some set and those that are not. The details of how this theory is worked out are too many and too technical for this little book. They would make another story.

Cantor's work deepened our intuition into the infinite, a concept that had challenged philosophers going back to ancient Greece. Russell, on the other hand, called attention to the errors into which our intuition may lead us.

Research on infinite sets continues, as mathematicians try to develop the basic assumptions about sets, from which all their properties would follow by pure logic.

CHAPTER EIGHT

Twins

Some strings of *a*'s, *b*'s, and *c*'s contain two shorter strings, or *words*, right next to each other that are identical. For example, the string *cbabcabcac* has two adjacent copies of *abc*. As another example, *baabc* has two neighboring copies of the short word *a* of length 1. On the other hand, the string

acbcabcbacb

has no such repetitions next to each other.

Call two identical words that are next to each other in a string *twins*. Perhaps "Siamese twins" may be more descriptive since the words are stuck together. A string that contains no twins we will call *twin free*.

How long can a string in the letters a, b, *and* c *be if it is twin free?*

This question and related ones were raised and answered by the Norwegian mathematician Axel Thue (1863–1922) in 1912. His motivation was simply the desire to know the answer. As he explained at the beginning of his paper, "For the development of the logical sciences it is important to find large areas for speculation about difficult problems without any consideration of possible applications."

However, the question was destined to be asked several times later. The American mathematician Marston Morse (1892–1977) raised it and related questions in 1921, when he used the answer to one of the questions in his study of the motion of a particle on certain curved surfaces. S. E. Arshon (a Russian mathematician who disappeared in a Stalinist reign of terror and died in prison around 1940), published his answer to the question stated above in 1937; he had heard the question at a conference in 1933. P. C. Novikov (1901–1975), another Russian mathematician, used Arshon's result in a paper published in 1959 that solved a famous problem in algebra raised by the English mathematician W. Burnside (1852–1927) in 1902. Morse didn't know about Thue's work, and Arshon had no idea that the question had already been answered by Thue and by Morse. The applications of Morse and Novikov are too technical to be described here.

The history of Thue's problem illustrates two recurrent facts about the practice of mathematics. First, discoveries made only to satisfy curiosity may many years later have totally unexpected applications. Second, several mathematicians, working independently, may ask the same question and come up with the same discovery.

Before we begin to consider Thue's question, we should pause to see what the answer is when we are permitted strings made up only of a's and b's, instead of a's, b's, and c's. His question then becomes, "How long can a twin-free string of a's and b's be?"

Such a string might as well start with an a. The next letter must be b; otherwise, we would form the twins aa. So we now have the string ab. The next letter must be a, and we reach aba, a string of length 3.

Can we add a fourth letter? We can't add an *a* since then we get an *aa*. But we can't add a *b* either, for then we would obtain *abab*, which consists of twins. We are blocked, and we conclude that the length of a twin-free string using just two letters cannot be greater than 3. Later in this chapter we will use the six twin-free strings on the letters *b* and *c*, which are *b*, *bc*, *bcb*, *c*, *cb*, and *cbc*.

Before Thue began his investigation he must have wondered, "Could I make very long twin-free strings? Or is there an upper limit to their length? After all, the longer a string grows, the more chance there is that twins will form and I'll be stopped."

Before reading further, you may want to stop and see how long a twin-free string you can devise on the letters *a*, *b*, and *c*. Even a few minutes experimenting will help you appreciate the reasoning in the rest of the chapter.

The Method

The many solutions of Thue's problem offered over the years all depend on a technique called *expansion*. This method involves the careful choice of three words, which I will denote *U*, *V*, and *W*. Each of them is some string made up of the letters *a*, *b*, and *c*. (The ones that we will use later in the chapter are *U = abcab*, *V = acabcb*, and *W = acbcacb*.)

These three words must have the following property: If one starts with a twin-free string and replaces each *a* in it by the string *U*, each *b* by the string *V*, and each *c* by the string *W*, then the new longer string formed is again twin free. This procedure can then be applied to the longer string to produce an even longer twin-free string. Applying the expansion repeatedly would yield twin-free strings as long as we please.

To illustrate the process, assume that *U* is *aba*, *V* is *cba*, and *W* is *bcb*. If we apply the process to the string *bacb*, we obtain the string

cba aba bcb cba   or   cbaababcbcba

The only catch is finding the right words *U*, *V*, and *W*. The choice in the example is not right since it replaces a twin-free string with one that has twins—in fact, four of them, *aa*, *abab*, *bcbc*, and *cbcb*.

My Search

I thought that before reading what various mathematicians had found, I should spend a little time trying to find words *U*, *V*, and *W* that work. If I were lucky and happened to find such a triple, then I wouldn't need to read what others had done. (Sometimes it's harder to understand someone else's solution than to find a solution yourself. In the first case you have to deal not only with the mathematical substance but also with the idiosyncrasies of another person's perspective and writing style.) If I didn't find such a triple, *U*, *V*, and *W*, at least I would be better prepared to read and appreciate the papers that describe three words that do the job.

I wanted to find three words *U*, *V*, and *W* such that when I replaced *a*, *b*, and *c* throughout any twin-free string by *U*, *V*, and *W*, respectively, I obtained another twin-free string. I will call the initial string the *original string* and the longer string obtained by the replacements, the *expansion*, or *expanded string*.

I looked for three short words. For convenience I narrowed my search to the case in which they all have the same lengths.

In the simplest case of interest, *U, V,* and *W* all have length 2. I decided to assume that $U = ab$, since at this point in the construction, *a, b,* and *c* all play the same interchangeable roles.

With this choice of *U,* both *V* and *W* must begin with *a* or *c* and end with *b* or *c.* (If, for instance, *V* begins with *b,* then *UV* contains the twins *bb.*) There are therefore four possibilities for *V* and for *W: ab, ac, cb,* and *cc.* Both *ab* and *cc* are immediately ruled out: We already have $U = ab$, and *cc* consists of twins. Thus *V* and *W* are each either *ac* or *cb.* Since there are only two possibilities, one is *V* and the other is *W.* I decided that I may as well assume that $V = ac$ and $W = cb$. Then $VW = accb$, which includes twins. This showed that *U, V,* and *W* cannot all have length 2.

I spent a few more minutes ruling out the case in which the three words all have length 3.

At that point I felt that the problem was certainly interesting, and I decided to do a little more experimenting. Rather than trying to find three suitable words of length 4, I tried some promising choices of *U, V,* and *W.* For instance,

$$U = ab, \qquad V = acb, \qquad \text{and} \qquad W = acabcb$$

None of these contains twins. Even better, none of the cases in which two of them are adjacent—*UV, UW, VU, VW, WU,* and *WV*—includes twins, as is easy to check. Unfortunately, *UVW* does, as it contains *bacbac:*

U	V	W
ab	acb	acabcb

I had done enough to see that finding a suitable *U, V,* and *W* is not a trivial task. Moreover, I had a better feeling for the problem and was ready to appreciate the solution. I knew what obstacles had to be overcome.

The Solution

The simplest solution I found appeared in a paper of P. A. B. Pleasants in 1970, in which he considered several related problems. This is his choice of three words:

$$U = abcab, \quad V = acabcb, \quad \text{and} \quad W = acbcacb$$

I will call these three particular words *blocks*. They have lengths 5, 6, and 7. As a result, when an expansion is made with their aid, the expansion will be more than five times as long as the original string.

Pleasants' argument doesn't even use arithmetic. Instead, it depends on common sense and attention to details. Follow the steps carefully. The proof is an instance of proof by meticulous bookkeeping.

Let us illustrate the procedure before we show that it always transforms a twin-free string to another twin-free string. We begin with any twin-free string. For the sake of simplicity, we choose the string *a* of length 1. Expansion replaces *a* by

$$U = abcab$$

Next, we repeat the procedure, so that *abcab* now plays the role of the original string. Replacing *a* by *U*, *b* by *V*, and *c* by *W* gives us

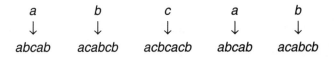

a	b	c	a	b
↓	↓	↓	↓	↓
abcab	acabcb	acbcacb	abcab	acabcb

It is not as easy to check that this expanded string of length 29 is twin free just by eyeballing it. Each of the twins could have any length from 1 to 14. Even a seemingly thorough search may fail to spot any twins (if there are any).

Expansion of our string of 29 symbols using *U*, *V*, and *W* would produce a string of more than 150 letters, much too long to be easily checked by eye. Showing that expanding a twin-free string with the aid of these blocks always produces another twin-free string requires careful attention to details.

A quick glance shows that the three blocks *U*, *V*, and *W* are twin free. That is encouraging. Also, the six cases formed of two different blocks—namely, *UV*, *UW*, *VU*, *VW*, *WU*, and *WV*—are also twin free. For instance, inspection shows that there are no twins in

$$UV = abcabacabcb$$

You may want to check the other five cases.

We must show that no matter how long an original twin-free string may be, the substitution of *U* for *a*, *V* for *b*, and *W* for *c* produces another twin-free string. Clearly, we must get some insight into those three blocks in order to establish this property. We can't write down all possible strings and check them by inspection.

Remarks on the Three Blocks

Each block begins with *a* and ends with *b*. Each block contains two *a*'s, a *front a* and a *rear a*. Each block is composed of two sections that begin with *a*, which we call a *front section* and a *rear section*.

All told, there are six sections. Each section consists of an *a* followed by one of the six twin-free words made with the letters *b* and *c*. These six words using *b* and *c* we call *tags*. There are three *front tags* in the front sections and three *rear tags* in the rear sections.

The six tags are all different. Front tags end in *c;* rear tags end in *b*. It follows that as you run through the *a*'s in the expansion of a twin-free string using these blocks, the letters that precede the *a*'s alternate . . . *b, c, b, c, b,* . . . .

It is easy to decide whether an *a* in an expansion is a front *a* or a rear *a*. Just look at the letter immediately left of it. If that letter is *b*, then the *a* is a front *a*. If that letter is *c*, then the *a* is a rear *a*.

Outline of the Proof

We will prove that the expansion of a twin-free string by Pleasants' three blocks yields another twin-free string. For convenience we will break the argument into three cases. In each case we will denote a pair of twins by TT^* (tee and tee star). T and T^* are twins, with T^* the same word as T. We give the twins different names in order to be able to distinguish them in the following discussion.

First, we will show that the expansion has no long twins T and T^*. By *long twin* we mean a twin in which at least two *a*'s appear. In this case, T, and hence T^*, is long enough to contain a complete tag. Second is the case in which T has exactly one *a*. Third, T has no *a*'s. Note that in the last two cases the twins are quite short. In the second case T has no more than six letters; in the third, no more than three letters.

The Proof

We begin the proof by analyzing the case where the twins are long. Assume that in the twins TT^*, T has at least two *a*'s.

First, consider the special case in which T consists of blocks. That implies that the first letter in T is a front a. T and T^* are made up of the same blocks in the same order. The following figure illustrates the typical case:

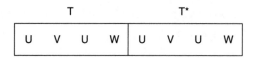

These twins would come from the expansion of the part of the original string shown here:

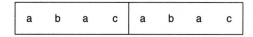

There are then twins in the original string. But since the original string has no twins, an arrangement such as that shown in the first figure cannot occur.

We see therefore that the expansion of a twin-free string has no twins TT^* such that T (hence T^*) consists of complete blocks.

The case of long twins TT^*, in which T does not consist of complete blocks, is more interesting. The twins in this case appear as shown in the next figure. It shows the case in which a twin contains two blocks; a twin could have any number of blocks.

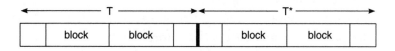

T begins with just part of a block. Thus T^* does too. That forces T to end with just part of a block, and hence T^* also.

We will show that *if there are twins of the type shown in this figure, then there are also twins of the type already considered, that is, twins that consist only of complete blocks.* That simpler case we have already shown is impossible.

Recall that T has at least two a's. We must examine the situation shown below, in which the first two a's in T (hence in T^* also) are shown.

There is a tag between the two a's shown in T. It ends in either b or c. Consider first the case in which it ends in b—that is, it is a rear tag. It is part of a block, denoted B in the next figure, that extends beyond T. The same block, denoted B^*, contains the left end of T^* and extends to its left into T.

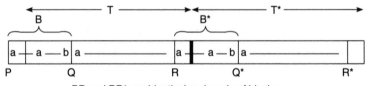

PR and RR* are identical and made of blocks

Consider the part of T not in B or B^*, labeled QR. It is made up of complete blocks. The same holds for the corresponding part of T^*, labeled Q^*R^*. QR and Q^*R^* are identical strings made up of the same blocks.

Inspection of this figure then shows that the string PR is identical to the string RR^*. Therefore they are twins. Moreover, they are composed of complete blocks. As we already saw, this cannot happen in the expansion of a twin-free string.

That takes care of the case in which the tag between the first two *a*'s ends in *b*. The case in which it ends in *c* is similar, and you may want to run through it. In that case the new twins formed, which consist of complete blocks, are shifted to the right instead of to the left.

We conclude that in an expansion of a twin-free string there are no long twins—that is, twins TT^* where T has at least two *a*'s.

No Short Twins

Two much easier cases remain, in which a twin has one *a* or none at all.

Take first the case in which T has exactly one *a*. Of course its twin, T^*, also has exactly one *a*, as shown:

The two *a*'s are much nearer each other than shown in the figure since at most three letters separate them.

If *a* is not the first letter in T, it is preceded by *b* or *c*. The second *a* is then preceded by *c* or *b*, respectively, since the letters preceding the *a*'s in a string made of blocks alternate. Thus T^* is not the same as T. It follows that *a* must be the first letter of T, and we have the arrangement shown here:

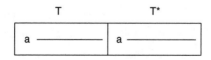

Could the two *a*'s come from just one of the blocks, $U = abcab$, $V = acabcb$, and $W = acbcacb$? It takes only a minute to rule out any such possibility. For instance, if they both came from *U*, the figure would look like this:

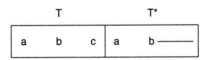

The letter to the right of *b* in T^* would be a front *a*, which violates the assumption that a twin has only one *a* (and the fact that T^* is identical to *T*). The blocks *V* and *W* can be ruled out just as quickly.

Could the *a*'s come from different blocks? In that case the *a* in *T* would be a rear *a* and the *a* in T^* would be a front *a*. *T* would be the back section of a block. The diagram is shown here:

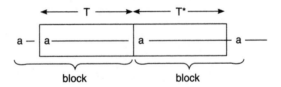

That means that the rear tag that appears in *T* must be an initial part of the front tag of a different block. Inspection of *U*, *V*, and *W* shows that this is never the case. In short, the case shown in this last figure cannot happen. This disposes of all cases in which a twin has exactly one *a*.

Finally, what if *T* has no *a*'s at all? The picture in this case looks like this:

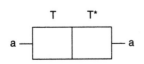

Again, the neighboring *a*'s in the figure are really much closer than they look since at most three letters separate them. *T* consists of at most one symbol, and so must T^*. Therefore TT^* is one of *aa, bb,* and *cc.* None of these occur because the blocks have no such twins, nor does a pair of adjacent blocks.

This completes Pleasants' reasoning and shows that the expansion of a twin-free string by the three chosen blocks is again twin free. Repeated expansions provide twin-free strings as long as we desire.

Questions

Typical of many mathematical discoveries, the proof we just finished raises still more questions. If ignorance is measured by the number of questions we can't answer, then, as we will see, we are more ignorant after the proof than before.

How many strings using the letters *a, b,* and *c,* of each given length, are twin free? Here is a table that shows the number of twin-free strings of each length up through length 24:

Length	Number of Strings	Length	Number of Strings
1	3	13	342
2	6	14	456
3	12	15	618
4	18	16	798
5	30	17	1044
6	42	18	1392
7	60	19	1830
8	78	20	2388
9	108	21	3180
10	144	22	4146
11	204	23	5418
12	264	24	7032

The numbers grow rapidly, but not wildly. Note how the ratios between successive numbers behave. For instance, we have, approximately, 4146/3180 = 1.304, 5418/4146 = 1.307, and 7032/5418 = 1.298. Perhaps, as more values are calculated, the ratios will approach a number near 1.3. If so, what is that number? Could it be 1.3 itself?

Is there a different way to construct long twin-free strings? A way that could be programmed for a computer? A way that would manufacture such strings letter by letter? I tried several approaches, described in Appendix B. One of them produced a string using four letters with length over 7 million, at which point I shut down the computer.

The computer presented data that suggest new questions and hunches, but it did not provide answers. However, without the aid of the computer, I wouldn't even have had data on which to base the questions. The computer is to mathematicians as a telescope is to astronomers or a microscope is to biologists. It magnifies the power of the researcher. It may exhibit a counterexample that refutes a theory, but it does not create theories or proofs.

A Bit of History

As I mentioned earlier, several mathematicians had solved Thue's problem, some unaware of earlier work, some aware but offering different solutions. Pleasants, whose paper contains several other results, was clearly unaware of the details of Thue's solution.

In an e-mail I asked Pleasants how he found his three words. "Just elementary messing about," was his reply. He continued:

> The problem came from a sheet of research problems circulated among Cambridge mathematicians. I knew the answer was known but didn't know what it was. Having

found it myself, I was encouraged to go on and try the related unsolved problem that was the main subject of my papers.

Redoing something that's known can draw you into a problem so that you can move on to something that's not known. If you just read what someone else has done, you think it's so impressive and complicated you couldn't possibly improve on it. Ignorance is a powerful tool for a mathematician. Maybe that's the advantage young mathematicians have.

After I understood Pleasants' solution, I thought it would be interesting to see how Thue solved the same problem in 1912. His solution, which is quite different from Pleasants', occupied a few pages hidden in a 67-page paper. It is sketched in Appendix B.

Imagine my surprise to see that Thue had used the very same blocks—*abcab, acabcb,* and *acbcacb*—that Pleasants had chosen. That both, independently, used the same three blocks suggests that these particular blocks may be the shortest possible. A. Capri in 1983 proved that they are indeed the shortest in the sense that their total length, $5 + 6 + 7 = 18$, is minimal.

However, the three blocks *abc, ac,* and *b* come close to transforming twin-free strings to twin-free strings, as M. Hall proved in 1964. If we always replace *a* by *abc, b* by *ac,* and *c* by *b,* and are careful in our choice of the initial twin-free string, these replacements will generate arbitrarily long twin-free strings. However, if we start with the twin-free string *aba,* those substitutions immediately produce a string with twins. They transform *aba* into *abcacabc,* which has twins.

If instead we start with the string *a,* those same substitutions repeated as often as we please will always produce twin-free strings. You could show this by reasoning much as we did in the case of long twins.

The last word on strings formed of letters has yet to be stated. The work continues, guided by new questions in various disciplines. For example, students of formal languages in theoretical computer science are examining strings that avoid patterns other than twins. Mathematical biologists are modeling the cellular growth of organisms such as seaweed with the aid of expansions in the spirit of those used in creating twin-free strings.

If, as it turned out, the history of the twin-free problem was often unknown even to those working on it, imagine how hard it is to predict its future and the future of related problems.

A Backward Glance

Now that we have completed our journey let us see where we have traveled.

In four of the chapters the strings are produced by chance: by throws of a needle (Chapter 1), by games played until a team leads by two points (Chapter 2), by streaks and slumps (Chapter 4), and by counting ballots (Chapter 6). These chapters sample the theories of probability and statistics, which are concerned with questions such as, "How many people should we poll to get a reliable estimate of what the entire population thinks?"

In two of these chapters (Chapters 2 and 4) we add more and more terms of an infinite collection of positive numbers. This illustrates the *limit process* of calculus. Calculus is the study of continuous change, such as the varying temperature throughout a lake or the trajectory of a rocket. Even the very definition of *temperature* requires calculus. Without calculus, we would be back in the eighteenth century, without trains, cars, planes, or electricity. Without electricity, there would be no radio, television, or computers.

Chapter 3 on the complete triangle comes from topology, the study of space. Because the chapter is concerned with triangles and polygons, we might think it is part of Euclidean geometry. However, the lines joining the dots can be wiggly curves; they do not have to be straight lines in order for us to carry out our counting of pebbles.

Near the end of that chapter we show that no matter how you use a given set of dots to cut a polygon into triangles, you always get the same number of triangles. The reasoning, which involves adding up angles in two ways, makes use of the fact that the angles of a triangle add up to 180 degrees. The conclusion remains true even if the edges are wiggly. In this more general case the proof must resort to topological concepts, for one no longer has the sum of the angles of a triangle as a tool.

Thrifty strings (Chapter 5) and Twins (Chapter 8) come from the field called *combinatorics,* the branch of mathematics concerned with finite sets. Combinatorics plays a big role in such varied applications as designing experiments, managing airline schedules and reservations, and computer programming. It is one of the few areas of mathematics in which an amateur can make a significant contribution.

Infinity (Chapter 7) is part of set theory, which provides the foundation of all mathematics. In contrast to combinatorics, it is concerned primarily with infinite sets.

The eight chapters offer a glimpse into just a few of the hundreds of branches of mathematics thriving today. We live in a golden age of mathematics, both pure and applied, an age that stretches back to the Renaissance without interruption. It shows every sign of going on indefinitely. Mathematics remains vital, stimulated by old unsolved problems and by new questions coming to it from such diverse sources as molecular biology, chemistry, physics, computing, and mathematics itself.

Mathematics can tempt all those who have preserved the spirit of the explorer.

Appendixes

Triangles

In Chapter 3 we used the fact that the sum of the three angles inside a triangle is 180 degrees—that is, it is a straight angle. The proof for this is short and elegant. After we present it, we generalize the theorem to polygons and apply the generalization to dissections of a polygon into triangles.

A bit of background.

In the figure below four angles are labeled x, z, y, and t. We will use the same letters to denote the size of each angle, as measured in degrees. We will show that x is equal to y.

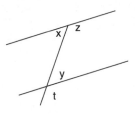

The two lines that look parallel actually are.

We have to start somewhere, and thus we will assume that z equals t, an assumption that the picture suggests is certainly plausible. Now $x + z$ is equal to $y + t$ since both angles in each sum add up to a straight angle. Because z equals t, it follows that x equals y.

Now consider a typical triangle, with angles A, B, and C, as shown here:

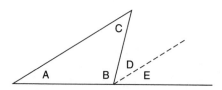

The dashed line is parallel to the side AC of the triangle. We will use this figure to show that $A + B + C = 180$ degrees. Observe that E equals A and that D equals C. Now,

$$B + D + E = 180 \text{ degrees}$$

Replacing D with C and E with A in this equation shows that $B + C + A$ equals 180 degrees also.

Knowing the sum of the angles in any triangle, we can easily find the sum of the angles in any polygonal circuit. For example, any quadrilateral can be cut into two triangles, as shown here:

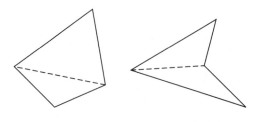

Hence the sum of the angles in a four-sided polygon is 2 times 180, or 360 degrees. Similar reasoning shows that the sum of the angles in a pentagon, or five-sided polygon, is 3 times 180, or 540 degrees.

More generally, the sum of the angles in any polygon, is (the number of sides − 2) times 180 degrees. This total can also be expressed as "(the number of corners − 2) times 180 degrees." In a moment we will use this second form.

In Chapter 3, we showed that no matter how a given set of dots is used to cut a polygon into triangles, the number of triangles will always be the same. Now we give an explicit formula for the number of those triangles. The angles in each triangle in a dissection contribute 180 degrees. Now count up the total of all the angles in terms of the dots, as we did before.

At each interior dot there are 360 degrees. At each dot on the boundary that is not a corner, there are 180 degrees.

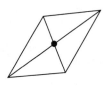

Triangles around
inner dot

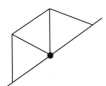

Triangles around
border dot

The sum of the angles at the corners is (the number of corner dots − 2) times 180 degrees. Therefore:

180 × number of triangles = (360 × number of interior dots)
+ (180 × number of border dots that are not at corners)
+ [180 × (number of corners − 2)]

Dividing both sides by 180 shows that

> Number of triangles = (2 × number of interior dots)
> + number of border dots + number of corners − 2

To put it more simply,

> The number of triangles is 2 less than the total number of dots plus the number of interior dots.

Once we choose the dots, this relation tells us how many triangles there will be, no matter how we choose to join the dots to form the triangles.

Twins: A Supplement

This appendix describes Thue's theorem and also my experience using the computer to make twin-free strings.

Thue's Theorem

Thue took a very general approach by asking: What are some conditions on three words U, V, and W such that substituting U for a, V for b, and W for c would replace any twin-free string in a, b, and c by a (longer) twin-free string? As we saw in Chapter 8, $U = abcab$, $V = acabcb$, and $W = acbcacb$ are three such words.

He found that the following two conditions are enough. The first should not surprise us; that is, if it were not satisfied, some twin-free strings of length 3 would be expanded to strings with twins.

1. Each of the following 12 strings, when written in terms of a, b, and c, is twin free: UVU, UVW, UWU,

UWV; VUV, VUW, VWU, VWV; WUV, WUW, WVU, WVW.

2. None of the three words—*U, V,* or *W*—appears in any of the other two words.

Thue's proof is similar to Pleasants' for the case of long twins.

If the three words have the same lengths, then condition 2 is automatically satisfied. Therefore, to check that three words of equal length expand twin-free strings to twin-free strings, all we must do is check whether any of the 12 mentioned strings is twin free.

The Computer and Twins

The approach used by Thue, Pleasants, and others didn't completely satisfy my curiosity. I hoped that there would be a more direct construction, one that builds a twin-free string one letter at a time. I decided to experiment to see if I could devise such a method.

I first tried several "automatic" approaches by hand. I began by building up a string letter by letter, always choosing the letter, *a, b,* or *c,* that occurs earliest in the alphabet and does not introduce twins into the string I was forming. That method gives me a start of *a.* At the next step it gives *b* since the choice of *a* would produce the twins *aa.* The next step provides another *a,* and I reach the string *aba.* At the next step, *a* is forbidden because *aa* would form. Moreover, *b* is also since it would produce the string *abab,* and twins again. So the next letter is *c,* and I reach *abac.* The next three steps add *aba,* and I have *abacaba.* Then the method runs into an obstacle. Both *a* and *b* clearly produce twins. But so does *c,* as you may check.

Always choosing the "earliest" gave me a twin-free string of length 7. Next I decided to try a mix of "earliest" and "latest." By "latest" I mean "choose the letter that occurs latest in the alphabet and does not lead to twins." Of course, I was still restricted to the first three letters of the alphabet.

Letting E stand for earliest and L for latest, I tried ELL, a pattern I repeated until I got stuck.

Starting with E, earliest, I got an a. Then L, latest, gave me c, and so I had ac. The next L gave me b since using c would produce twins, cc. At that point I had obtained acb. Then I went on with ELL again. The next E gave me an a, and I had $acba$. L then produced $acbac$. The next L led to an a because both c and b introduced twins.

Continuing with this pattern yielded a string of length 26:

> *ELL ELL ELL ELL ELL ELL ELL ELL EL*
> acb aca bcb acb cac bac abc acb ac

Adding a to this string produces twins TT^* with T of length 8; adding b produces twins with T of length 3; adding c produces cc.

I then tried other combinations of E's and L's using a program that the mathematician Dean Hickerson wrote for the computer. With each attempt the program stopped. That raised a new question: Will every combination of E's and L's, when applied to three letters, always stop? I don't know, and, as is usual in mathematics, if you can't answer a question, you ask what you hope is an easier one.

Will every combination of E's and L's, repeatedly applied to four letters, always stop?

Using Hickerson's program, which could apply to any number of letters, I experimented. Several combinations that we tried stopped. However, the direction *EEEEEEL*, which stands for "six earliest followed by one latest repeatedly applied" led to a very long string. When we shut down the computer after three solid days, the string had over 7 million letters. We don't know whether the computation would ever have stopped.

It would require a book of some three thousand pages to record the twin-free string the machine calculated. I wonder how long it would have taken to produce that string letter by letter by hand, and then how long to check it.

To give some idea of the computer's speed, I should mention that the first thousand letters were produced in a split second, almost as soon as my fingers left the keyboard. However, as the string lengthened, the computer slowed down.

In any case I pose this very concrete question: Does the program with *EEEEEEL* (six *E*'s and an *L*) ever stop when applied to four letters?

We might expect that a program that goes on for 7 million steps will go on forever, but that assumption may not be valid. Consider what happened when I tried *EEEEELLLL-LLL* (five *E*'s followed by seven *L*'s). After it produced a twin-free string with 100,000 letters, I felt that the machine would never stop. Indeed, it passed 200,000, then 300,000 letters. It was finally blocked when the string had grown to 357,665 letters.

Clearly, the computer raises new questions but, fortunately, does not answer them.

For Further Reading

The following books contain many chapters accessible to the lay reader

R. Courant and H. Robbins, *What Is Mathematics?*, 2d ed.; rev. by I. Stewart. Oxford University Press, New York, 1941, reprinted.

T. Dantzig, *Number, the Language of Science*. 4th ed., Free Press (Simon and Schuster), New York, 1985.

K. Devlin, *All the Math That's Fit to Print*. Mathematical Association of America, Washington, D.C., 1994 (143 short articles, almost all published in the *Manchester Guardian*).

D. Gale, *Tracking the Automatic Ant*. Springer, New York, 1998 (columns from *Scientific American*).

M. Gardner, *Aha! Gotcha*. W. H. Freeman, New York, 1982.

M. Gardner, *Life and Other Mathematical Amusements*. W. H. Freeman, New York, 1983.

M. Gardner, *Penrose Tiles to Trapdoor Ciphers*. W. H. Freeman, New York, 1989.

M. Gardner, *Time Travel and Other Mathematical Bewilderments*. W. H. Freeman, New York, 1987.

S. Golomb, *Polyominoes*, Princeton University Press, Princeton, 1994.

S. K. Stein, *Strength in Numbers.* John Wiley & Sons, New York, 1996.

S. K. Stein, *Mathematics, the Man-Made Universe.* Dover, Mineola, 1997.

S. K. Stein, *Archimedes: What Did He Do Besides Cry Eureka?* Mathematics Association of America, Washington, D.C., 1999.

The following technical references are either sources for some of the chapters or take some of the ideas further.

Chapter 1   The Needle and the Noodle

R. E. Miles and J. Serra, *Geometrical Probability and Biological Studies,* Lecture Notes in Biomathematics, 23d ed. Springer-Verlag, New York, 1978.

Chapter 2   Win by Two

T. Gilovich, R. Vallone, and A. Tversky, The Hot Hand in Basketball: On the Misperception of Random Sequences. *Cognitive Psychology,* 17 (1985): 295–314.

S. K. Stein, Existence Out of Chaos. In R. Honsberger (ed.), *Mathematical Plums,* Mathematical Association of America, Washington, D.C., 1979, pp. 62–93. *(This has several examples of adding a collection of numbers in two ways to obtain information.)*

Chapter 3   The Complete Triangle

These references illustrate some of the uses of Sperner's lemma.

R. Hochberg, C. McDiarmid, and M. Saks, On the Bandwidth of Triangulated Triangles, *Discrete Mathematics,* 138 (1995): 261–265.

E. A. Kasimatis, Dissections of Regular Polygons into Triangles of Equal Areas, *Discrete and Computational Geometry,* 4 (1989): 375–381.

P. Monsky, On Dividing a Square into Triangles, *American Mathematical Monthly,* 77 (1970): 160–164.

E. I. Sperner, Neuer Beweis für die Invarianz der Dimensionzahl und des Gebietes, *Abhandlungen aus dem Mathematischen Seminar der Hamburgischen Universität,* 6 (1928): 265–272.

E. I. Sperner, Fifty Years of Further Development of a Combinatorial Lemma, pts. A and B. In W. Forster (ed.), *Numerical Solutions of Highly Nonlinear Problems.* North Holland Publications Co., 1980, pp. 183–214.

S. K. Stein, A Generalized Conjecture about Cutting a Polygon into Triangles of Equal Areas, *Discrete Computational Geometry,* 24 (2000): 141–145.

S. Stein and S. Szabó, *Algebra and Tiling,* Mathematics Association of America, Washington, D.C., 1994, pp. 107–133.

F. E. Su, Rental Harmony: Sperner's Lemma in Fair Division, *American Mathematical Monthly,* 106 (1999): 930–942.

S. Wagon, Fourteen Proofs of a Result about Tiling a Rectangle, *American Mathematical Monthly,* 94 (1987): 601–617.

Chapter 5 Thrifty Strings

J. Burns and C. J. Mitchell, Coding Schemes for Two-Dimensional Position Sensing. In M. J. Ganley (ed.), *Cryptography and Coding 3,* Clarendon Press, Oxford, 1993, pp. 31–66.

S. W. Golomb, *Shift Register Sequences.* Holden-Day, San Francisco, 1967. *(This describes how to produce thrifty strings algebraically.)*

Chapter 6 Counting Ballots

W. Feller, *An Introduction to Probability Theory and Its Applications,* Vol. 1, John Wiley & Sons, New York, 1950, chap. 3.

Chapter 8 Twins

A. Lindenmayer, Mathematical Models for Cellular Interactions in Development, *Journal of Theoretical Biology,* 18 (1968): 280–315.

P. A. B. Pleasants, Non-repetitive Sequences, *Proceedings of the Cambridge Philosophical Society,* 68 (1970): 267–274.

P. Prusinkiewicz and A. Lindenmayer, *The Algorithmic Beauty of Plants.* Springer-Verlag, New York, 1990. *(This contains many pictures of objects produced by algorithms that generalize the expansion technique of Thue and Pleasants.)*

G. Rozenberg and A. Salomaa (eds.), Lindenmayer Systems: Impacts on Theoretical Computer Science, *Computer Graphics, and Developmental Biology.* Springer-Verlag, New York, 1992.

Subject Index

Index

About the Author

Sherman Stein received his Ph.D. from Columbia University. After a one-year instructorship at Princeton University, he joined the faculty at the University of California, Davis, where he taught until 1993. His main mathematical interests are in algebra, combinatorics, and pedagogy. He has been the recipient two MAA awards: the Lester R. Ford Award for Mathematical Exposition, and the Beckenbach Book Prize for *Algebra and Tiling* (with Sándor Szabó). He has also received The Distinguished Teaching Award from the University of California, Davis, and an Honorary Doctor of Humane Letters from Marietta College.